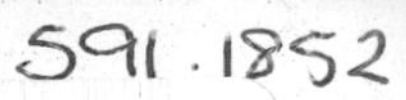

Understanding Muscles

Understanding Muscles

A practical guide to muscle function

Bernard Kingston

MA (Oxon), PGCE, DO, MRO
Lecturer in Functional Anatomy,
The British School of Osteopathy, London, UK
Registered Osteopath

CHAPMAN & HALL MEDICAL
London · Weinheim · New York · Tokyo · Melbourne · Madras

Published by Chapman & Hall, 2–6 Boundary Row, London SE1 8HN, UK

Chapman & Hall, 2–6 Boundary Row, London SE1 8HN, UK

Chapman & Hall GmbH, Pappelallee 3, 69469 Weinheim, Germany

Chapman & Hall USA, 115 Fifth Avenue, New York NY 10003, USA

Chapman & Hall Japan, ITP-Japan, Kyowa Building, 3F, 2-2-1 Hirakawacho, Chiyoda-ku, Tokyo 102, Japan

Chapman & Hall Australia, 102 Dodds Street, South Melbourne, Victoria 3205, Australia

Chapman & Hall India, R. Seshadri, 32 Second Main Road, CIT East, Madras 600 035, India

Distributed in the USA and Canada by Singular Publishing Group Inc., 4284 41st Street, San Diego, California 92105

First edition 1996

Typeset in 10/12 Univers by Type Study, Scarborough

Printed in Great Britain by The Alden Press, Osney Mead, Oxford

ISBN 0 412 60170 2 1 56593 123 8 (USA)

A catalogue record for this book is available from the British Library

Library of Congress Catalog Card Number: 95-71375

♾ Printed on permanent acid-free text paper, manufactured in accordance with ANSI/NISO Z39.48-1992 and ANSI/NISO Z39.48-1984 (Permanence of Paper)

To my parents, Charles and Helen, my family, my friends and Maxine, for their love, support and wisdom

‘What a piece of work is a man!’

William Shakespeare
Hamlet: Act 2, Scene 2

Contents

Preface ix

Acknowledgements xi

Illustration acknowledgements xiii

1 Introduction: how to use this book 1

2 Anatomical terminology and movements 5

3 Muscles and movement 9

4 The shoulder 21

5 The elbow 45

6 The wrist 59

7 The hand 71

8 The hip and pelvis 95

9 The knee 125

10 The ankle 141

11 The foot 153

12 The respiratory (and abdominal) muscles 167

13 The muscles of the vertebral column 187

14 The temporomandibular joint 215

Glossary of terms and abbreviations 227

References 235

Index 237

Preface

In this book I am seeking to encourage interactive learning. The urge to write it sprang from a feeling of frustration, first as a student and then as a lecturer, that in spite of the many excellent anatomy textbooks on the market, it remained a very difficult task to learn and remember the functional anatomy of muscles.

As a student, I longed for a book on muscles containing skeletal outlines on which the muscles could be drawn and studied individually. As a lecturer, I therefore began to produce notes which included skeletal outlines requiring muscles to be drawn in by students. In my experience this approach improves students' confidence and understanding of how muscles produce their particular movements.

This book makes no special claim to increase physiological understanding of muscle or fascia. It simply seeks to make life easier for people who find it difficult to understand why muscles produce the movements that they do, and would like an explanation centred on the joints over which muscles work. The often-puzzling classical names of muscles are also explained.

I have also found from experience that the addition of active muscle function tests is a valuable way of reinforcing the understanding of muscle attachments and function. The study of each muscle is accompanied by interactive study tasks which include muscle tests where appropriate, as well as instructions and general questions. It is suggested that students practise and observe these tests on themselves or willing colleagues as appropriate, but sensible rules of health and safety should be observed at all times. It is my belief that none of the tests are dangerous when performed responsibly, but undue force which might lead to injury should be avoided. If in doubt, avoid the muscle tests, or seek qualified advice first. Neither the author nor the publisher can be held responsible for any injury sustained while performing any tasks described in this book.

It is difficult when writing a practical book on muscles to know what to leave out. This book deals only with voluntary skeletal muscle. Furthermore, it is deliberately concerned with those muscles which act on the main joints of the body. The reason for this is that the interactive approach which is adopted here tends to work best if the student can shade lines between attachment points passing over bony levers. This includes the majority of muscles in the human body. I felt that this approach might be less suitable for the more intricate muscles of the urogenital region, as well as the face, eye, tongue, throat and so on. I am keeping an open mind on this, and will consider the possibility of remedying this deficiency in the future.

However, for the moment the book does not deal with every muscle in the human body, and I apologise for any irritation that this may cause. The fault lies entirely with my own judgement on the practical scope of this particular work. Where omissions have occurred, I hope nevertheless that the reader will be encouraged to look at muscles in a different and more interactive way, and perhaps adopt the investigative approach which is used here.

Finally, if the reader is able to gain as much insight and enjoyment from the reading as I have from the writing, then the task will undoubtedly have been worthwhile. I hope that this is the book that you have been waiting for.

Bernard Kingston

Acknowledgements

I should like to express my thanks and gratitude to the following people for their help and support either as friends, colleagues, teachers or students.

John and Dinah Badcock; Sally Champion; Simon Curtis; Jonathan Curtis-Lake; Dr Martin Collins; Maxine Dadson; Gabrielle Dowdney; Derrick Edwards; Lisa Fraley; Laurie Hartman; Alison Jesnick; Brian Joseph; Robin Kirk; Alain Lebret; Robin Lynam; Stephen Lusty; Lynne Maddock; Peter Mangan; Rosemary Morris; Hubert and Diana Moore; Doreen Ramage; Stephen Sandler; Christopher R.A. Smith; Steen Steffensen; Catherine Walker; Brad Wilson; the staff at Chapman & Hall; and finally, all my past and present students at the British School of Osteopathy for their unwavering friendliness and good humour. Without them, this book would not have been possible.

Illustration Acknowledgements

We would like to thank the following publishers for permission to adapt material. We made every effort to contact and acknowledge copyright holders, but if any errors have been made we would be happy to correct them at a later printing.

Green, J.H. and Silver, P.H.S. (1981) *An Introduction to Human Anatomy,* Oxford University Press, Oxford.

Hertling, D. and Kessler, R. (1990) *Management of Common Musculoskeletal Disorders* 2nd edn, J.B. Lippincott Company, Philadelphia.

Kapandji, I.A. (1987) *The Physiology of the Joints,* Churchill Livingstone, London.

Kapit, W. and Elson, L. (1977) *The Anatomy Coloring Book,* Harper and Row.

Snell, R.S. (1995) *Clinical Anatomy for Medical Students* 5th edn, Little, Brown and Company, Boston.

Williams and Warwick (eds) (1980) *Gray's Anatomy* 36th edn, Churchill Livingstone, London.

Chapter 1

Introduction: How to use this book

The aim of this book is to encourage you to discover how and why certain muscles move the body in the way that they do. You will be encouraged to participate in the learning process by shading lines between the designated muscle attachments, so that the dynamic action of the muscle is appreciated.

It is my belief, based on my experience as a teacher, that students learn the functions of muscles more effectively when they interact with a multi-dimensional diagram. All too often students are to be found gazing at an illustration where the details of a muscle (and often a large number of muscles) have been added. This is neither interactive nor stimulating, and the student does not find the muscle functions easy to learn or comprehend.

This book adopts a different approach. Only the attachment points of an individual muscle are shown on a skeletal outline. The essential muscle details are given for reference purposes, and this is followed by a number of practical **study tasks**. In each case you will be encouraged to 'learn by doing' and participate in the following activities:

- Shading lines between the attachment points indicated, and considering the muscle functions which are listed in the text.
- Either self-testing and/or observing the muscle functions on your own body, or a colleague as appropriate. 'Colleague' can of course be taken to mean anyone who is suitable and willing, but obviously frail or otherwise unsuitable subjects should not be asked to participate, for proper reasons of safety. The normal rules of professional conduct and commonsense should apply.

Each muscle studied is put in the context of the joint over which it operates. Therefore, the main chapter headings relate to particular areas of the body, peripheral joints and so on, except for the chapter on respiratory and abdominal muscles. The order of study does not matter particularly, but you are advised to start with the introductory chapters, and to revise or learn the basic anatomical principles and terminology. It is assumed that most readers will possess a basic knowledge of anatomy, or are following a related course.

I suggest that you study each muscle in the following way:

Step 1: Read the introductory section of the chapter and familiarize yourself with the anatomy of the appropriate joint.

Step 2: Study the details of each movement and muscle given, including the explanation of the name of the muscle.

Step 3: Carry out the study tasks, of which the key task will usually be to shade lines between the attachment points and consider the muscle functions. You should then imagine what happens when the attachment points are drawn closer together by muscle contraction. In most cases the functions will be clear.

I suggest that the labels of diagrams should be highlighted, and it may be helpful to recite the terms (quietly, if appropriate) in order to reinforce the learning process. Wherever possible, the labels are numbered and it is the number that appears on the diagram, which should help revision and self-testing.

Most introductory study tasks will encourage reference to a bone specimen. A plastic model will often suffice, but increasingly scarce bone specimens are usually more satisfactory, and you should obtain these if possible.

The main details of each muscle are usually given under the heading of the main movement that it produces (e.g. **flexion** of the shoulder joint). This is preceded by an illustrated explanation of the relevant movement, which will show the approximate active range of movement in the joint concerned, as well as listing the muscles producing that movement. Page numbers are given so that an individual muscle can be referenced in the text. If a muscle has already been described in detail in an earlier context, the page number will facilitate cross-referencing. **It is therefore particularly important to consult the illustrations of gross movements when seeking a complete list of muscles producing particular movements. The text will not include a full list, since an individual muscle may have been described fully in an earlier chapter.**

Example (from Chapter 5, 'The elbow')

Brachialis ('muscle of the arm')

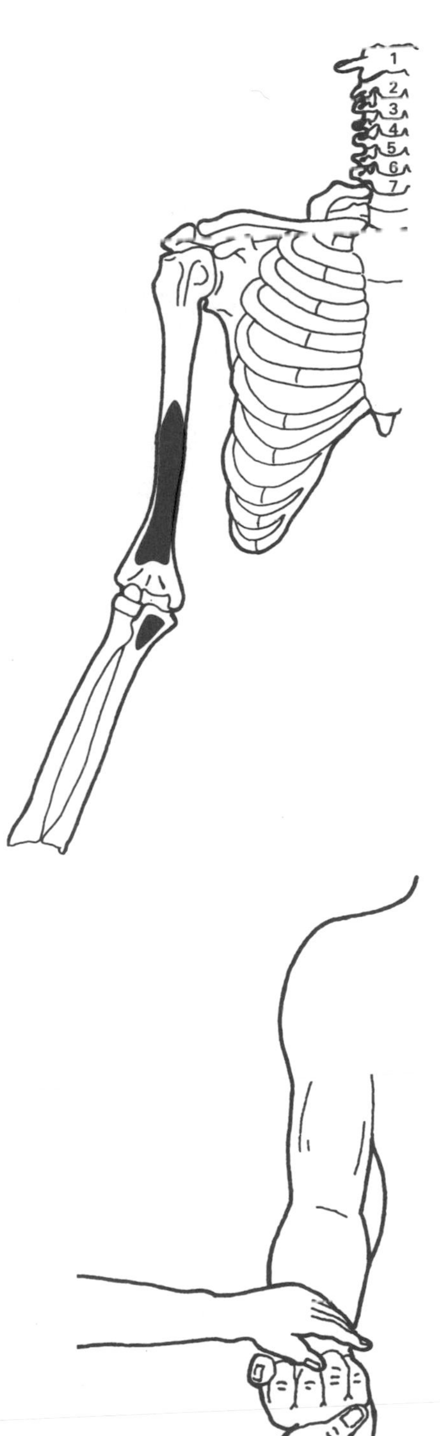

Attachments

- Lower part of anterior surface of the humerus.
- Below the coronoid process of the ulna.

Nerve Supply

Musculocutaneous nerve (C5, C6): the lateral portion is partly supplied by the radial nerve (C7).

Function

- Flexion of the elbow joint

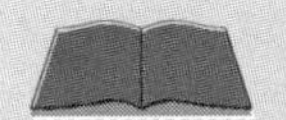

Study tasks

- Shade lines between the attachment points and consider the muscle function.
- Test the function of brachialis (given with biceps in the actual text) on a colleague with elbow flexed to 90° and palm facing upwards forming a loose fist. Resist at the wrist as they attempt to flex their elbow.

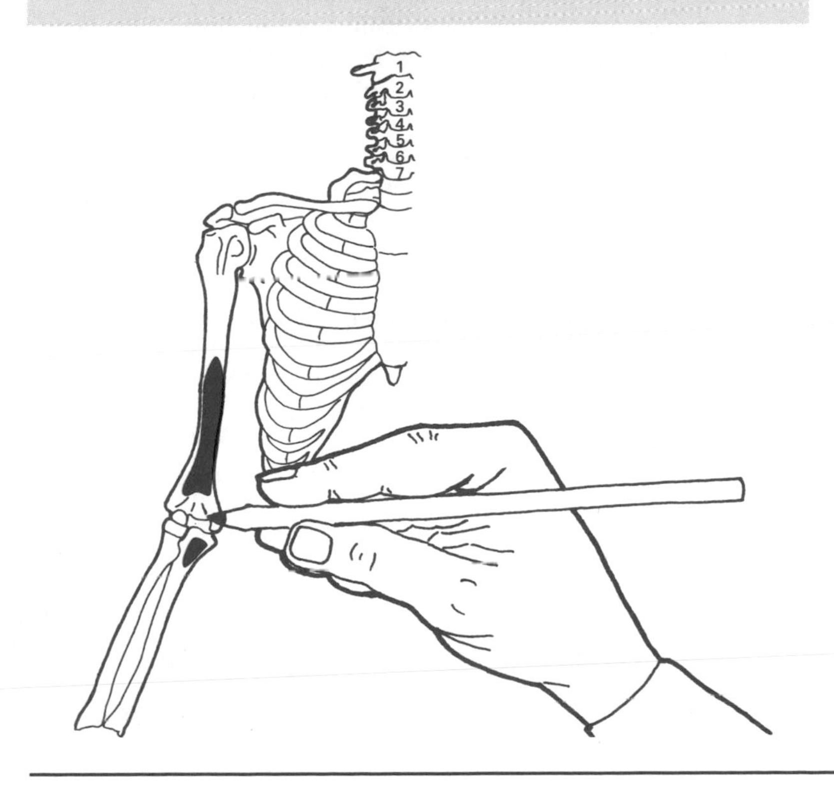

Worked example

Step 1: Read the introduction to the elbow joints (pp. 45–47).
Step 2: Study the details of the muscle as given.
Step 3: Perform the study tasks for this muscle.

Points to note

1. Use coloured pencils or felt tips depending on your artistic skills. It may be advisable to practise initially with a pencil which can be easily erased. Try to draw the boundaries of the muscle first, and be prepared to consult an atlas of anatomy if you are uncertain about the actual shape of an individual muscle (e.g. *Grant's Atlas of Anatomy*, 1991).
2. The colours used are a matter of taste, but it may be helpful to adopt a consistent approach, such as shading muscles in red and ligaments in green, wherever possible.
3. Sometimes the attachment of a muscle appears on a concealed bone surface. In this case dotted lines will be used, accompanied by an explanatory note.
4. The instruction to highlight labels, names of muscles and ligaments, bone features and so on, can be taken to mean any form of visual emphasis: underlining, especially in colour, is a perfectly satisfactory alternative.
5. Some muscles appear under different headings. The main details will usually be given in the context of the muscle's primary function, and full repetition of details is avoided if possible.
6. All joints illustrated are synovial unless stated otherwise. The type of synovial joint is indicated as appropriate.
7. The skeletal outlines usually portray only a half-view of the human body. The muscles will of course be present on the other side as well!
8. For the sake of clarity, avoid shading the numbers on certain diagrams.
9. All values relating to ranges of movement are approximate.

Chapter 2

Anatomical terminology and movements

The terminology used in this book to describe muscles and their movements conforms to that used in standard anatomy texts.

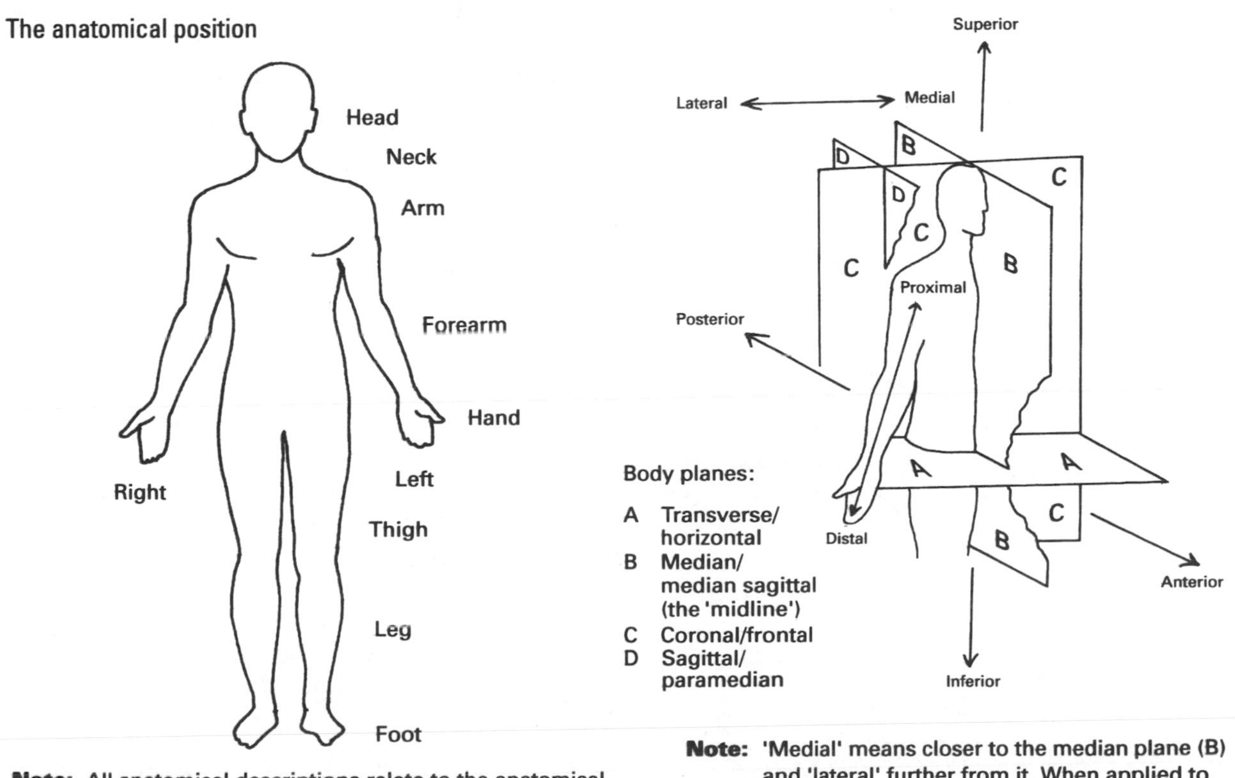

Note: All anatomical descriptions relate to the anatomical position shown above. The conventional terms used to describe the limbs are also shown.

Note: 'Medial' means closer to the median plane (B) and 'lateral' further from it. When applied to rotation movements, the terms 'internal' (medial) and 'external' (lateral) may be used.

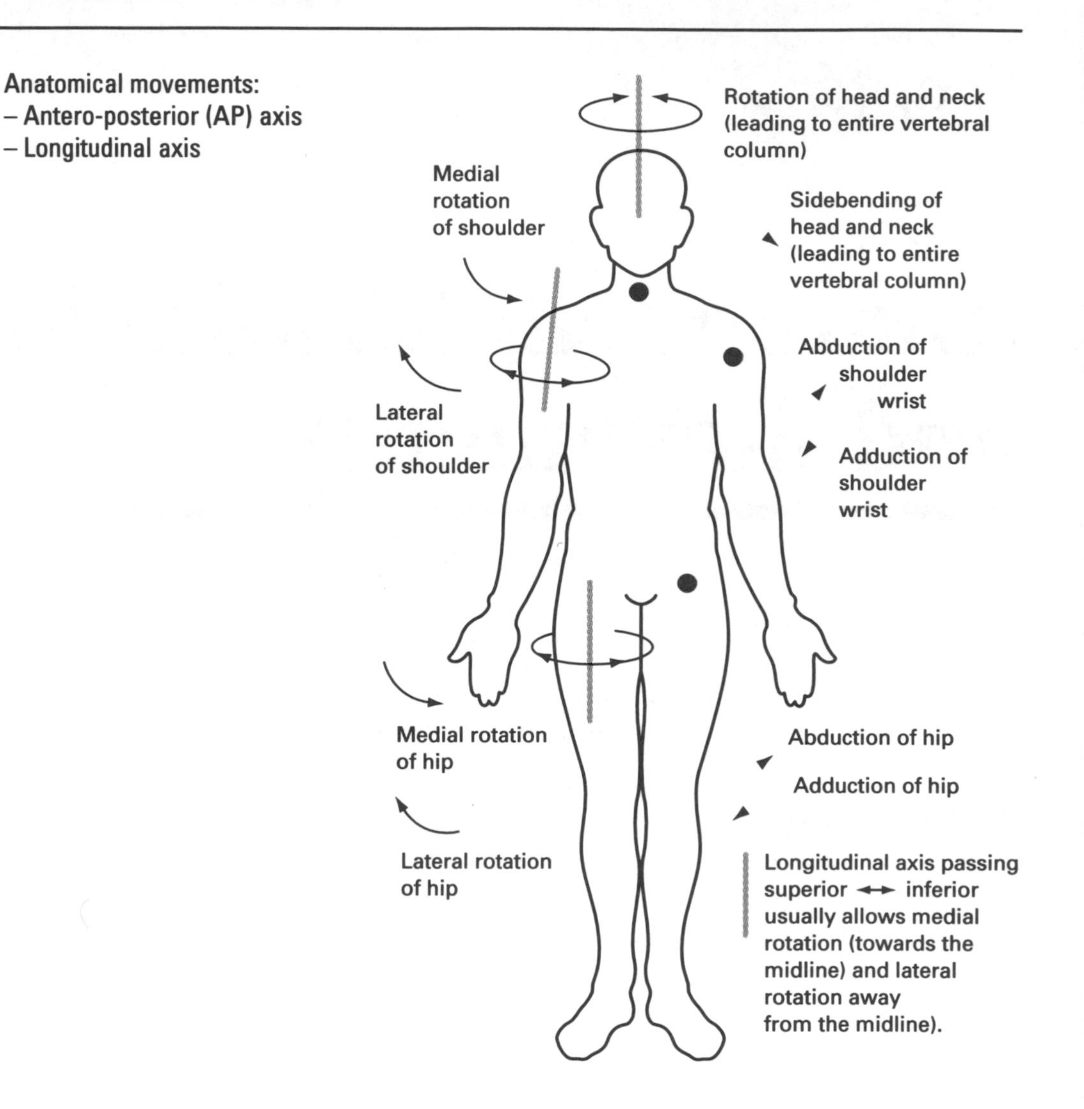

● Antero-posterior (AP) axis passing front ↔ back through the body allows abduction (movement away from the midline) and adduction (return to the midline).

Note: Only a sample of the main axes are shown.

A number of specialized movements are also found which will be described in the text as appropriate:

- pronation/supination
- inversion/eversion
- dorsiflexion/plantar flexion
- opposition
- gliding
- protraction/retraction
- depression/elevation
- circumduction

Anatomical movements: transverse axis

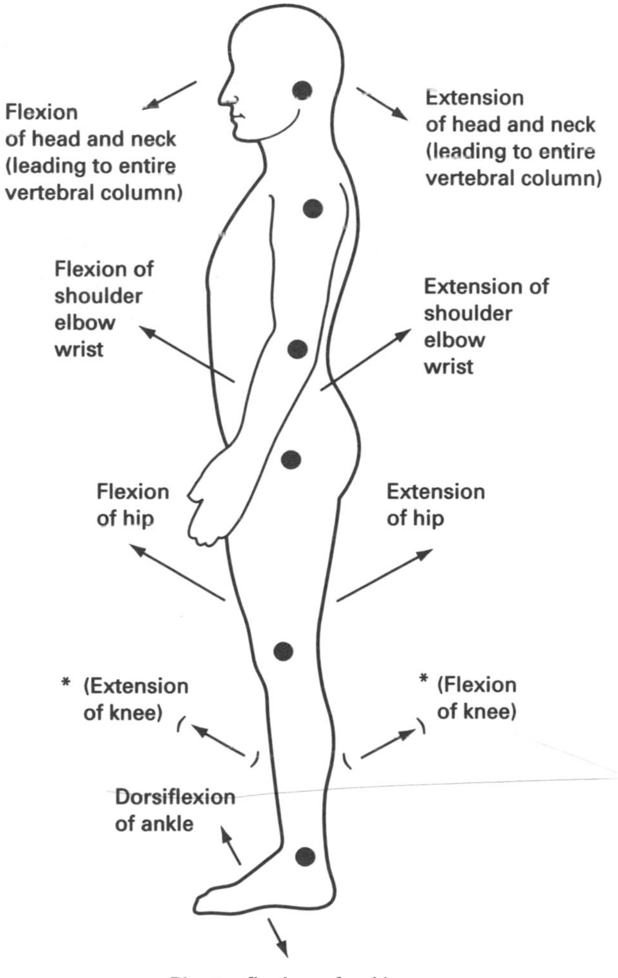

● Transverse axis passing left ↔ right through the body allows flexion (approximating anterior surfaces) and extension (approximating posterior surfaces).

Note: Only a sample of the main transverse axes are shown.

* For explanation of this anomaly see p. 131.

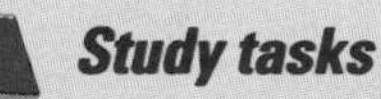

Study tasks

- Highlight the main details and features in each diagram.
- Shade the body planes in separate colours (p. 5).

Chapter 3

Muscles and movement

Introduction

There are three types of muscle in the human body, classified according to location, structure and nerve supply. The heart and gut are supplied by two separate types of involuntary muscle, but the muscles which are described in this book are all under voluntary control. They consist of bundles of fibres which contract to move bony levers, and therefore produce skeletal movement. The muscles under consideration here fall under the general classification of **voluntary striated skeletal muscle.**

Voluntary striated skeletal muscle

The basic structure of each muscle consists of bundles of muscle cells (fibres) which are bound together and enclosed by connective tissue. The resulting bundle is often referred to as the muscle **belly**. The **architecture** varies, but the basic structure of a typical **fusiform** (lozenge-shaped) muscle, such as biceps brachii, is shown below.

The structure of voluntary striated skeletal muscle

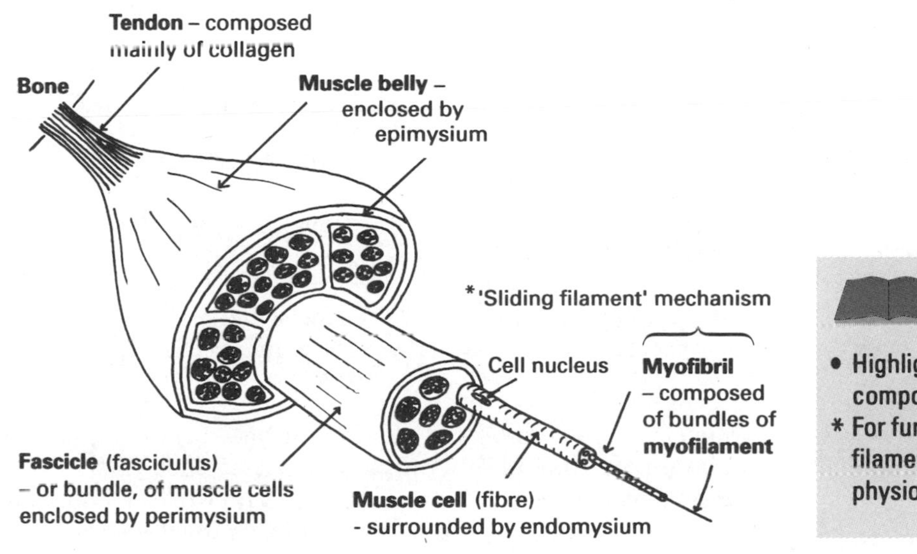

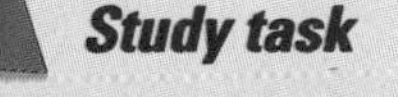

Study task

- Highlight the names of the tissue components shown.
- * For further details of the sliding filament mechanism consult a physiology text as appropriate.

Muscle attachments

In order to produce skeletal movement, a muscle generally attaches to bone. Most muscles taper to form a **tendon** composed of tough collagenous tissue, although some insert into flat tendinous sheets (e.g. over the abdomen) known as **aponeuroses**, or interlace with each other as **raphes**.

Types of muscle attachment

Simple single tendon attachment
• brachialis (p. 48)

Common tendon attachment of two or more muscle 'heads'
• gastrocnemius (p. 147)

Calcaneus bone

Aponeurosis – muscle fibres attach to flat tendon sheet

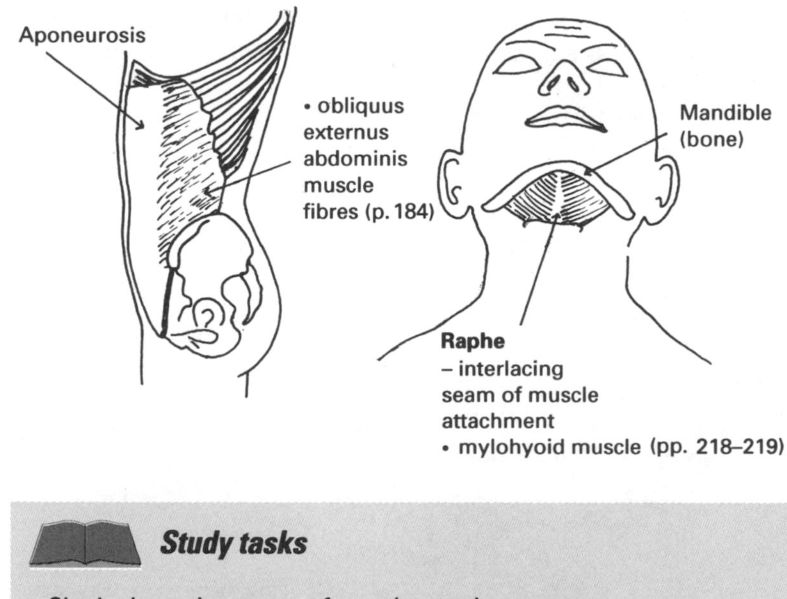

Study tasks

- Shade the various types of muscle attachment.
- Highlight the names and examples given, for future reference.

Muscles usually produce movement by crossing a joint or articulation, so that when contraction takes place one articulating bone is drawn towards another, providing leverage. The elbow joint provides a good example of this mechanism.

The elbow hinge joint: flexion and extension

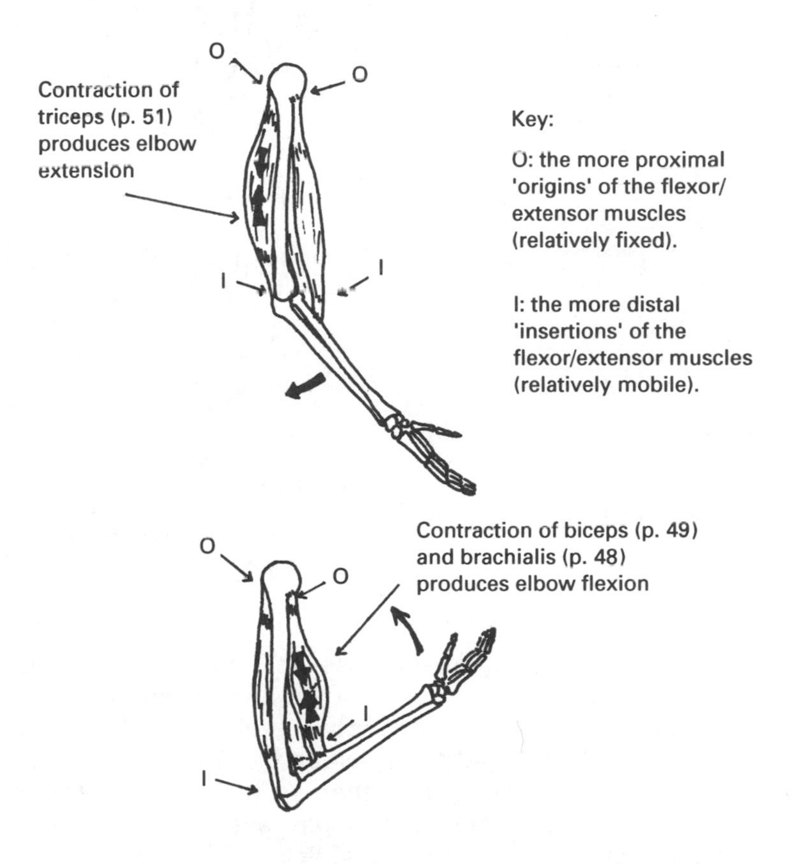

In order to produce these movements it can be seen that the attachment point at one end of the muscle is relatively immobile (sometimes called the **origin**), while the point of attachment at the other end produces the actual movement, and is often referred to as the **insertion**. Unfortunately, these terms can sometimes cause confusion, since the roles of the origin and insertion can be reversed in the same muscle.

For example, consider the functions of pectoralis major, a shoulder/thoracic muscle (full details pp. 32–33). It is attached to the sternum and ribs, but it also attaches to the humerus and produces important shoulder movements. If the rib attachments are regarded as the origin and the humerus as the insertion, the upper arm and shoulder may be moved variously into flexion, adduction and medial rotation (for illustration of these terms see pp. 6–7). However, pectoralis major can also be utilized as an accessory respiratory muscle of deep inspiration. In this case the arms may be raised (e.g. with hands clasped behind the head) and the humerus forms the relatively fixed origin, while the thorax is drawn upwards into inspiration by the insertion of the muscle into the ribs.

Pectoralis major

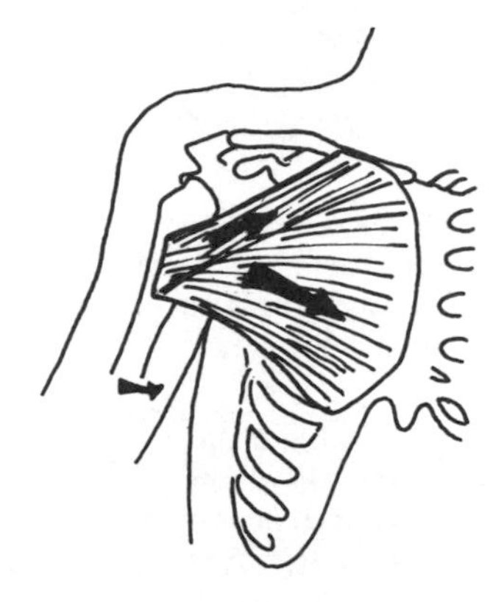

If clavicular and pectoral attachment points act as 'origins', the 'insertion' point into the humerus may adduct and medially rotate the shoulder.

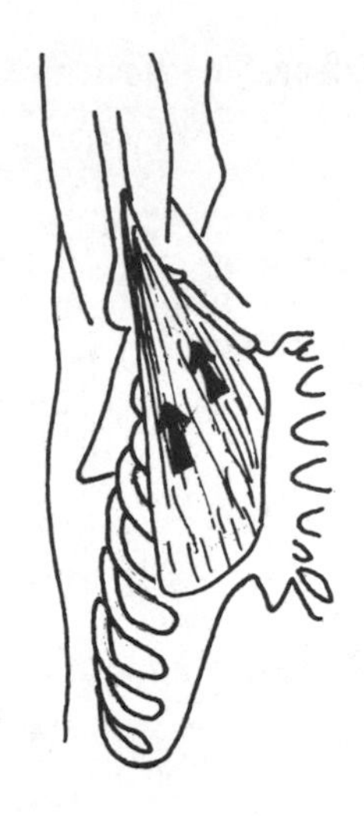

If the arm is raised and fixed the attachment point on the humerus in effect becomes the'origin'. The upper ribs may then be raised in deep inspiration.

While the terms 'origin' and 'insertion' are useful conceptually, their slavish use is best avoided, and for the purposes of this book the term **attachment** is preferred throughout. The reader is encouraged to consider the functions of individual muscles in context, and then to decide if the terms 'origin' and 'insertion' are appropriate, and in certain cases whether or not the terms are reversible.

Teamwork

Several muscles are usually recruited in order to produce effective movement. In elbow flexion (see above), the muscles which produce the main movement (brachialis and biceps brachii) are regarded as the **prime movers** or **agonists**. Muscles which produce the opposite movement (extension) must relax to allow flexion to take place. The main extensor muscle of the elbow is triceps brachii, and in this situation it is referred to as an **antagonist**. However, if extension of the elbow is required, the roles are reversed, making triceps the prime mover of extension, with biceps and brachialis as antagonists.

Further 'teamwork' may be necessary in realistic situations where muscles are used to move body parts. Consider the use of elbow flexion to lift a heavy tray. Steady controlled movement will require a whole range of muscles to stabilize the shoulder girdle in order to allow the elbow to perform its task. The 'extra' assisting muscles recruited to help are often termed **synergists** or **fixators**, literally fixing the body so that the prime movers can work more effectively.

Muscle architecture

The striated nature of muscle results from the sliding filament structure (p. 9), but the bundles of fibres or fasciculi are represented in differing forms depending on the type of function required. It is worth noting that function in the body usually does determine structure. In the case of muscle there is often a need for either increased **range** of movement (i.e. the need to move a bony lever some considerable distance), or **power** (i.e. if the muscle needs to be strong).

If range of movement is the main function required, the resultant structure may be a relatively long, but not necessarily powerful muscle.

If power is needed, the resultant muscle structure may be more compact, but may operate over a smaller range.

Some muscles show certain features of both, and are in effect a compromise between the two.

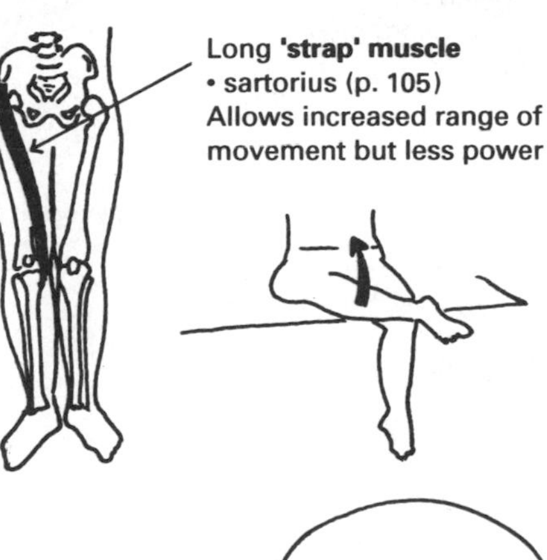

Shorter **quadrilateral** arrangement
• masseter (p. 221)
Allows increased power but smaller range of movement

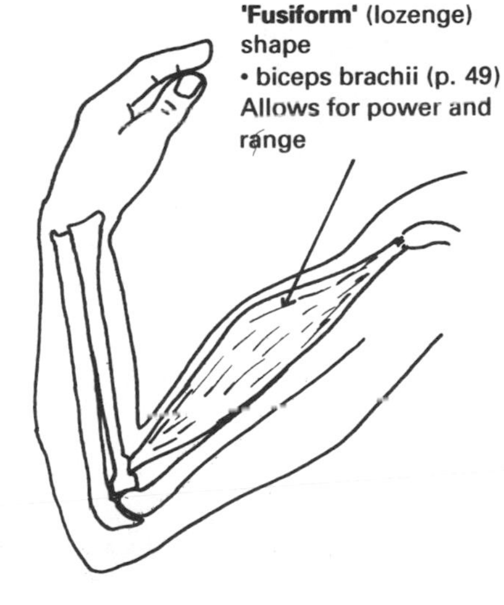

Another common presentation is the **pennate** type of arrangement, which is 'feathery' in detailed appearance, and seems to offer strength without bulk. These muscles usually need to be quite flat, or are found operating within confined spaces.

A triangular or convergent arrangement often occurs if the muscle attachment has to be focused or concentrated at a restricted point of leverage: and sphincters or body orifices require a circular arrangement of muscle fibres.

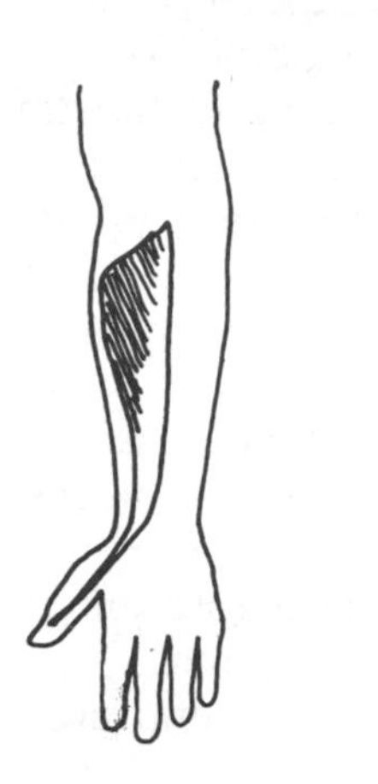

Unipennate
• flexor pollicis longus (p. 77)

Bipennate
• rectus femoris (p. 134)

Multipennate
• middle fibres of deltoid (p. 32)

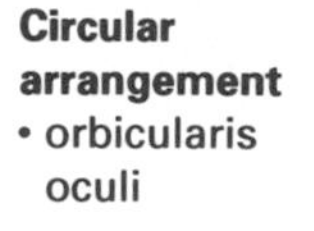

Circular arrangement
• orbicularis oculi

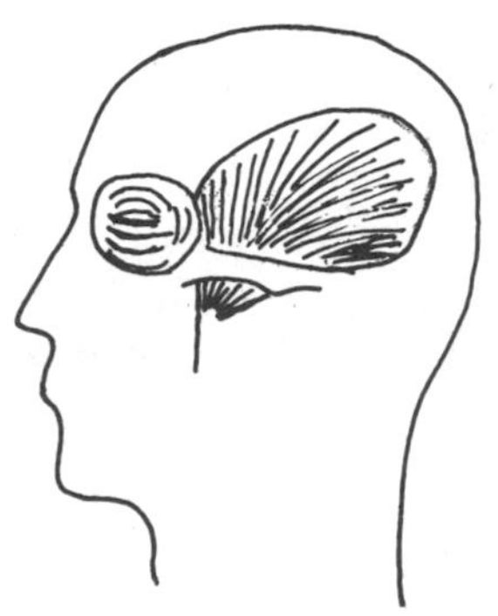

'Triangular' convergent arrangement
• temporalis (p. 222)

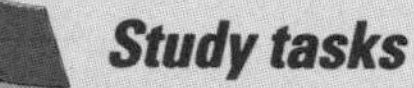

Study tasks

- Shade the different types of muscle arrangements in colour.
- Highlight the names of the features shown and examples given.

Different kinds and types of joints

Most movements take place using bones as levers, and where two articulating bones meet there will be a joint which allows a certain kind of movement to take place. An illustrated list of the main kinds of joints is given here, but the muscular movements described in this book usually operate over the various types of synovial joints in the appendicular skeleton, and the secondary cartilaginous joints in the axial skeleton (i.e. the vertebral column).

1. Fibrous joints (synarthroses)

These are found where only limited movement is desirable (slight 'give').
Articulating bone surfaces are held together by fibrous connective tissue

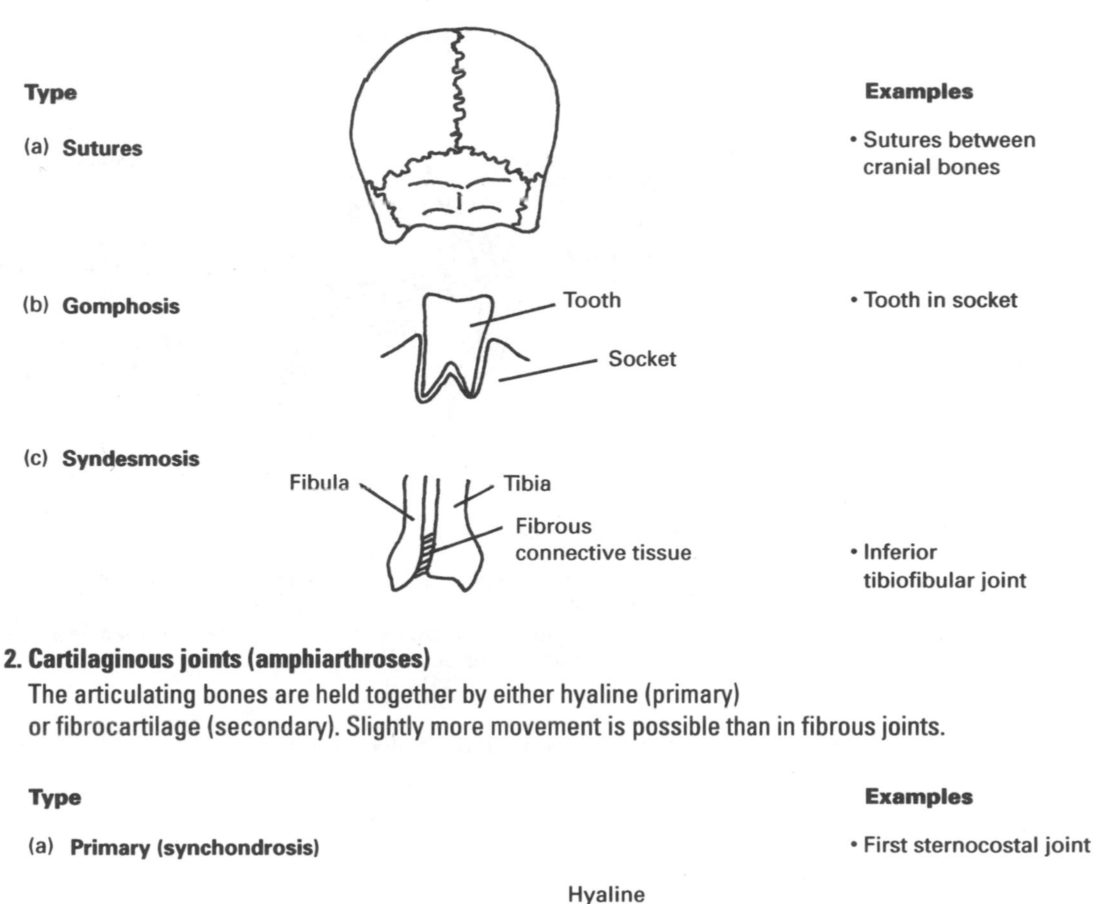

2. Cartilaginous joints (amphiarthroses)

The articulating bones are held together by either hyaline (primary) or fibrocartilage (secondary). Slightly more movement is possible than in fibrous joints.

Type	Examples
(a) **Primary (synchondrosis)**	• First sternocostal joint

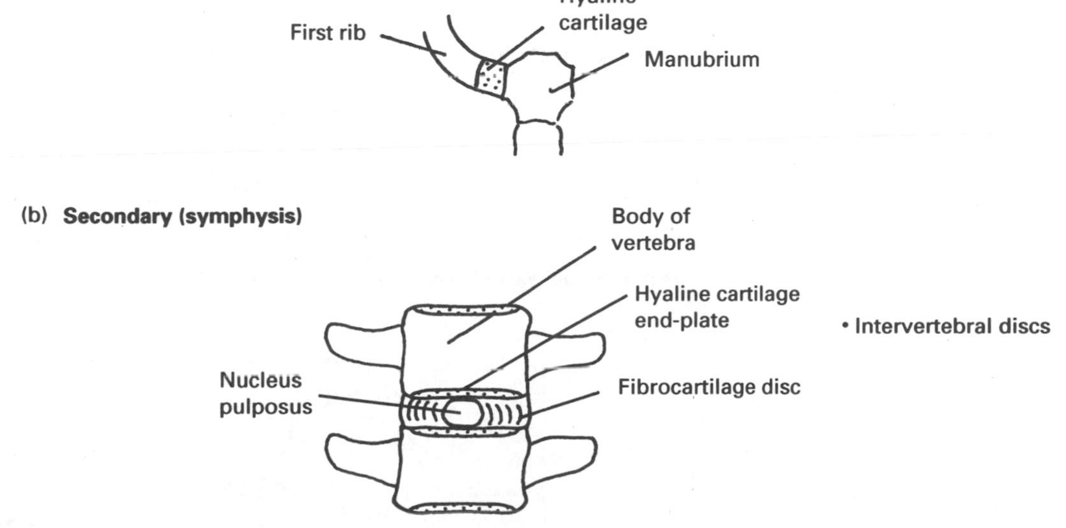

3. Synovial joints (diarthroses)

The presence of a joint cavity lubricated by synovial fluid allows a greater degree of movement. A fibrocartilage disc is sometimes present.

Generalized structure

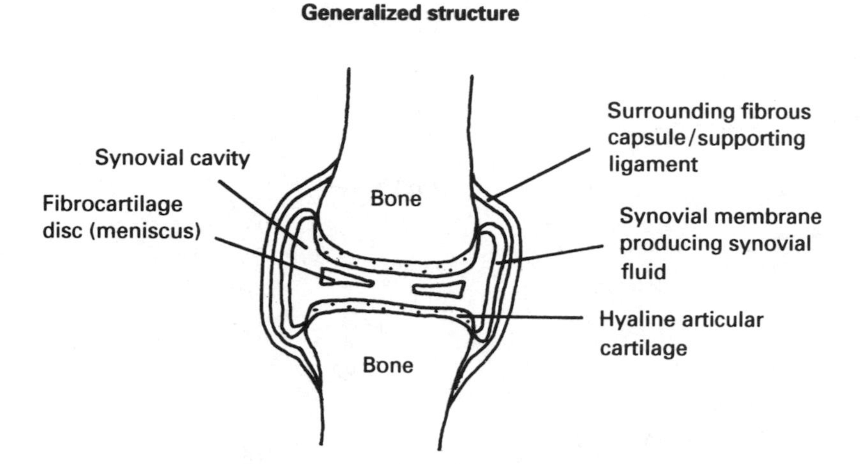

Types of synovial joint

Movements in synovial joints take place through either one axis (uniaxial), two (biaxial) or three (multiaxial). The directions of the axes are shown by arrows in the examples below. Imagine an axis as the 'axle' of a wheel.

A **transverse axis** allows **flexion/extension**

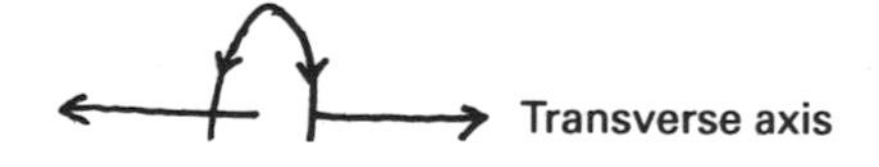

An **antero-posterior (AP) axis** (front to back) allows **abduction/adduction**

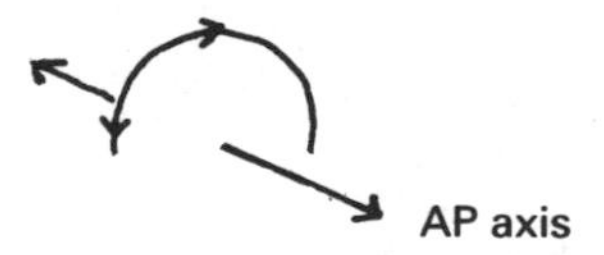

A **longitudinal axis** allows **rotation** movements

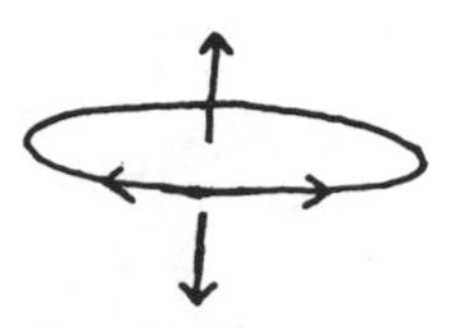

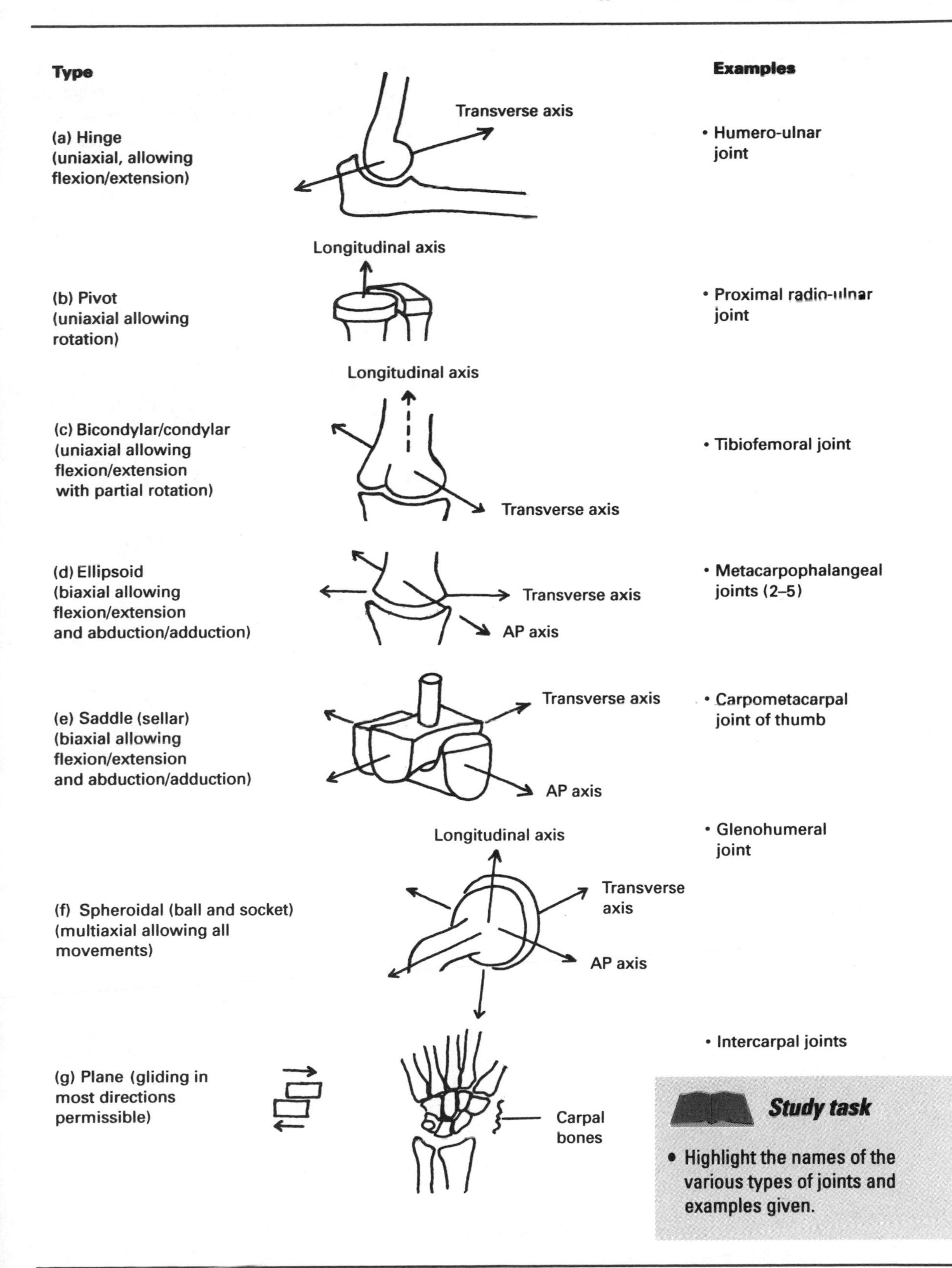

Study task

- Highlight the names of the various types of joints and examples given.

Levers

A lever is a rigid rod which rests on, and acts over, a relatively fixed point known as a fulcrum. In the human body, the bones act as the (relatively) rigid rods, and three classes of lever are usually recognized. The first class lever (see-saw arrangement) is a useful way of explaining why people strain their backs when lifting, because the see-saw model is a familiar one. The idea of a fixed 'effort arm' controlled by the back muscles straining to work over a vulnerable fulcrum (intervertebral discs), with a highly variable 'resistance arm', is a graphic way of advising that objects should be lifted close to the body, with a 'straight' back. 'Effort arm' is the distance between 'E' and 'F'; 'resistance arm' is the distance between 'R' and 'F'.

The first class lever

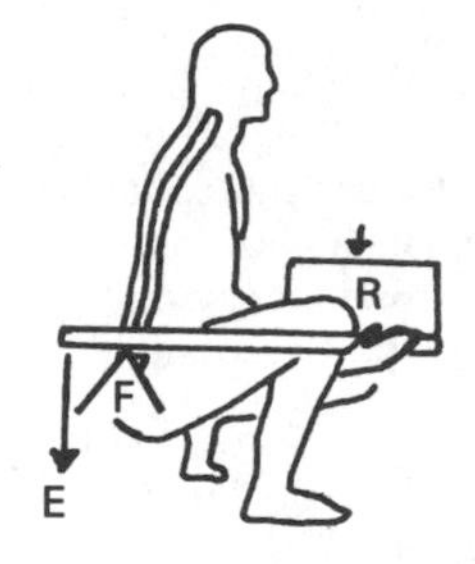

The second class lever

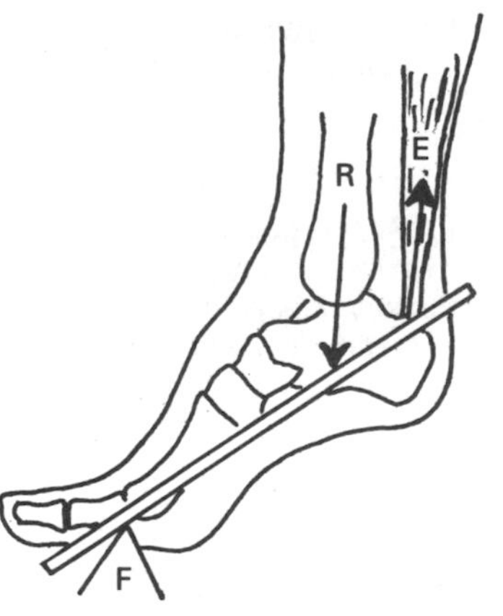

The third class lever

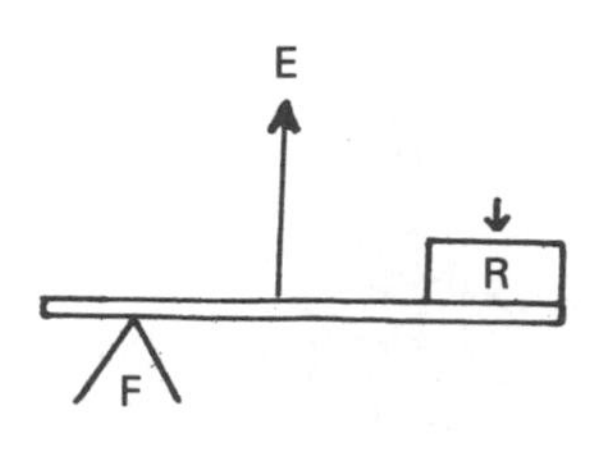

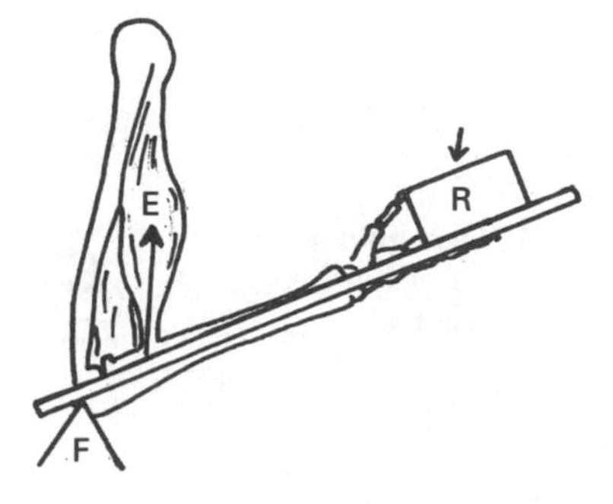

F= fulcrum
E= effort
R= resistance (load)

Study tasks

- Study the examples of the three types of lever and consider the implications for the human body.
- Consider the implications of good/bad lifting. Which structures and tissues are particularly vulnerable?
- Consider the effects of weak/strong back muscles, obesity, pregnancy and disc disorders, with reference to the first class lever.

Muscle names and functions

Muscles are generally named from Latin or classical Greek derivations, and an attempt is made in this book to provide a working translation, where the main details of the muscle are given.

There are also a number of principles at work, which may be useful to bear in mind, and the designated muscle characteristics are often combined.

Characteristic of muscle	Examples
Size	Teres major, teres minor
Location	Tibialis anterior, pectoralis major
Direction	Obliquus externus abdominis
Number of attachments	Biceps (2), triceps (3)
Attachment points	Sternocleidomastoid
Shape	Deltoid, trapezius
Position	Scalenus anterior, scalenus posterior
Function	Adductor magnus, flexor pollicis longus

Some principal functions of muscles

Functional name	Definition
Flexor	Generally brings anterior surfaces closer
Extensor	Generally brings posterior surfaces closer
Abductor	Moves body parts away from the midline
Adductor	Moves body parts closer to the midline
Rotator	Rotates body parts around a longitudinal axis
Supinator	Turns palm of hand anteriorly
Pronator	Turns palm of hand posteriorly
Sphincter	Enables orifices to open and close
Levator	Produces elevation or upward movement
Depressor	Produces downward movement
Tensor	Produces a greater degree of tension

Chapter 4

The shoulder

Introduction

The shoulder consists of more than one articulation. The main joint comprises the ball-shaped head of the humerus which articulates with the shallow glenoid surface (socket) on the scapula. The freedom of movement thus gained allows the hand and arm to move freely and perform a wide range of functions.

The scapula (shoulder blade) increases the scope of this movement still further by sliding over the rib cage under strict muscular control.

The clavicle (collar bone) attaches medially to the sternum and laterally to the acromion, and provides support anteriorly as an S-shaped strut which swivels as the rest of the shoulder moves.

Hence, it is appropriate to regard the shoulder as a girdle of interacting bones, muscles and joints.

The shoulder also acts as a social signal of expression, for example when we shrug our shoulders or let them droop; and they also form a pleasingly rounded contour when the surrounding muscles are well formed.

The bones and joints

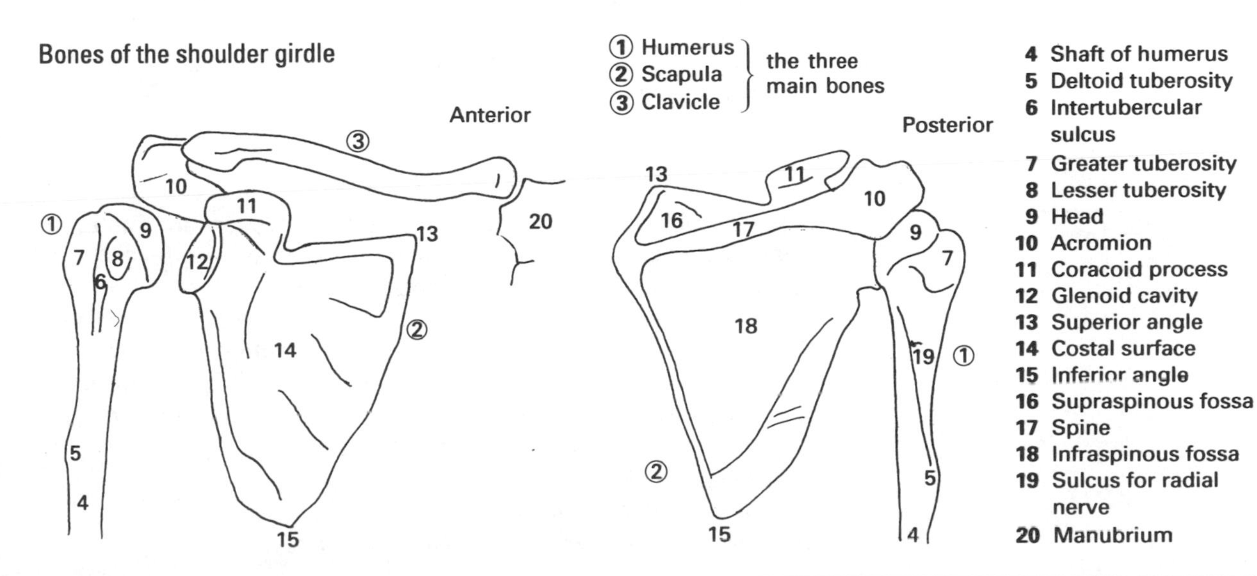

Study tasks

- Identify and shade the three main bones in separate colours.
- Highlight the names of the features indicated.
- Obtain bone specimens and identify the features shown.

Joints and ligaments of the shoulder girdle

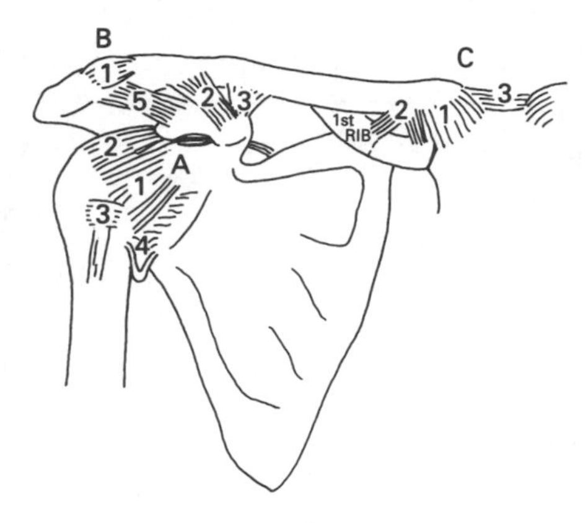

A Glenohumeral joint

Ball and socket.
Ligaments:
- **1** glenohumeral
- **2** coracohumeral
- **3** transverse
- **4** capsule and dependent pouch
- **5** (coracoacromial)

B Acromioclavicular joint

Plane (contains fibrocartilage meniscus).
Ligaments:
- **1** acromioclavicular
- **2**, **3** } coracoclavicular < trapezoid (2), conoid (3)

C Sternoclavicular joint

Modified saddle (contains fibrocartilage meniscus).
Ligaments:
- **1** anterior and capsule
- **2** costoclavicular (anterior and posterior)
- **3** interclavicular

Study tasks

- Study the details of each joint shown, and shade the ligaments.
- Highlight the names of the ligaments.

The muscles

The obvious role of muscles is to produce movement, but in the case of the relatively vulnerable shoulder there are further vital functions of protection and support. To some extent all muscles perform these functions, but the shoulder is particularly in need of protection and support, due to a combination of the shallow glenohumeral joint with the consequent mobility permitted, and the exposure to impact injuries. The muscles which provide specialist functions can be subdivided into:

- The superficial muscles
- The deeper musculotendinous rotator cuff.

The superficial muscles

The superficial muscles: anterior aspect

1 Anterior deltoid (p. 32)
2 Pectoralis major (pp. 32–33)
3 Trapezius (upper fibres) (p. 27)
4 Sternocleidomastoid (p. 173)
5 Latissimus dorsi (p. 37)
6 Serratus anterior (p. 28)
7 Biceps brachii (p. 34)
8 Biceps aponeurosis
9 Brachialis (p. 48)
10 Triceps brachii (medial head) (p. 51)
11 Triceps brachii (lateral head) (p. 51)
12 Triceps brachii (long head) (p. 51)
(13) Obliquus externus abdominis (p. 184)

The superficial muscles: posterior aspect

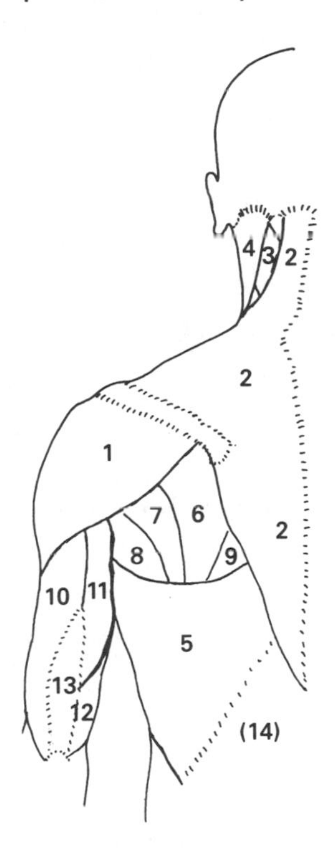

1 Posterior deltoid (p. 32)
2 Trapezius (p. 27)
3 Splenius capitis (p. 195)
4 Sternocleidomastoid (p. 173)
5 Latissimus dorsi (p. 37)
6 Infraspinatus (p. 42)
7 Teres minor (p. 42)
8 Teres major (p. 36)
9 Rhomboid major (p. 29)
10 Triceps brachii (lateral head) (p. 51)
11 Triceps brachii (long head) (p. 51)
12 Triceps brachii (medial head) (p. 51)
13 Triceps brachii tendon
(14) Thoracolumbar fascia

Study tasks

- Shade the muscles in colour.
- Highlight the names of the muscles.

The deeper musculotendinous rotator cuff

These muscles and their tendons converge onto the head of the humerus like fingers grasping a gearstick. This has a stabilizing effect, and helps prevent dislocation of the shallow gleno-humeral joint.

The rotator cuff muscles

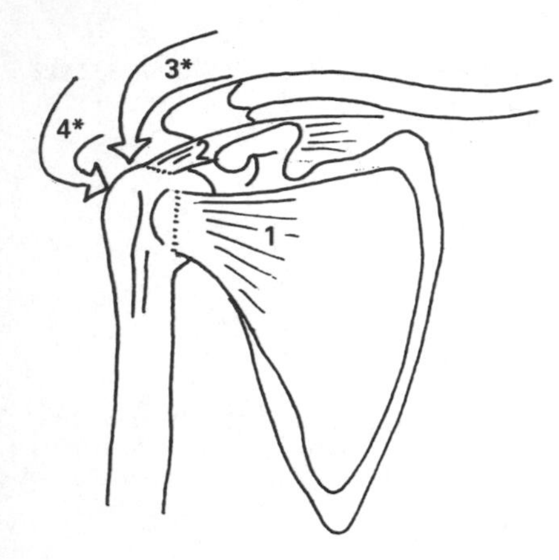

1 Subscapularis (p. 41)
2 Supraspinatus (p. 39)
3 Infraspinatus (p. 42)
4 Teres minor (p. 42)
* Schematic representation only

Study tasks

- Shade the muscles in colour.
- Highlight the names of the muscles.

Movements of the scapula

Movements at the shoulder demonstrate the coordination between the scapula, which is capable of moving independently, and the glenohumeral joint, whose range is increased by the accompanying movements of both the scapula and clavicle.

Elevation and depression

Elevation of the scapula occurs when shrugging the shoulders; and depression occurs either with gravity when the shoulders are relaxed, or actively, as when using crutches or gymnastically supporting the body on parallel bars.

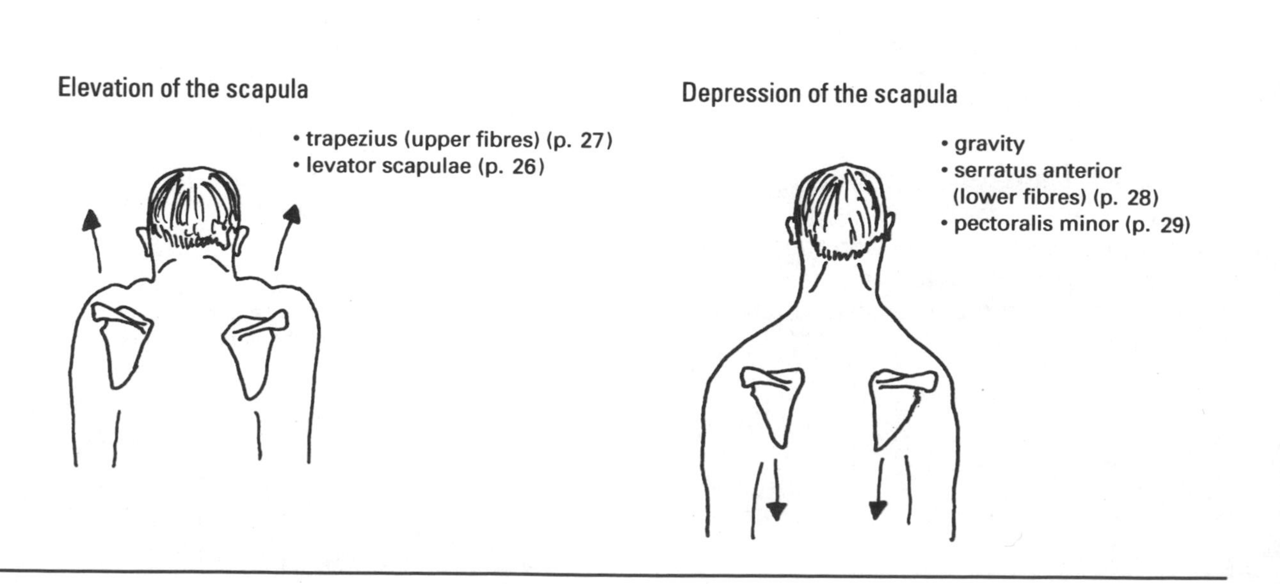

Protraction and retraction

These movements are produced either when punching and pushing (protraction) or when bracing the shoulders back (retraction).

Protraction of the scapula

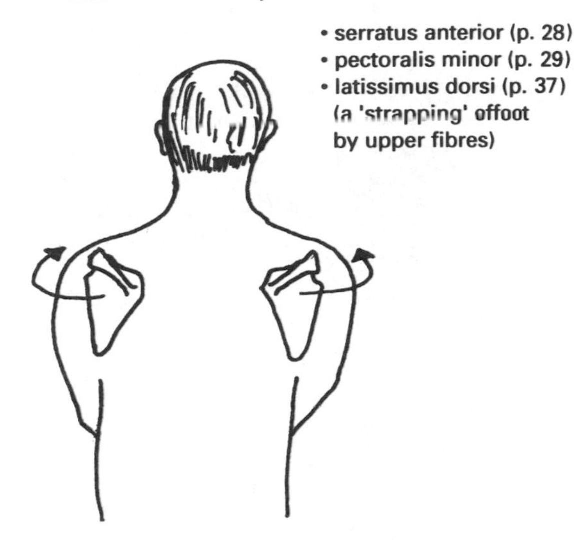

Retraction of the scapula

- trapezius (middle/lower fibres) (p. 27)
- rhomboids (p. 29)

Forward and backward rotation

The forward rotation of the scapula turns the glenoid cavity upwards and allows the arm to be raised above the head. Backward rotation is usually a return to the anatomical position by gradual relaxation of the forward rotator muscles, but can be produced actively.

Forward rotation of the scapula

- trapezius (upper fibres) (p. 27)
- serratus anterior (lower fibres) (p. 28)

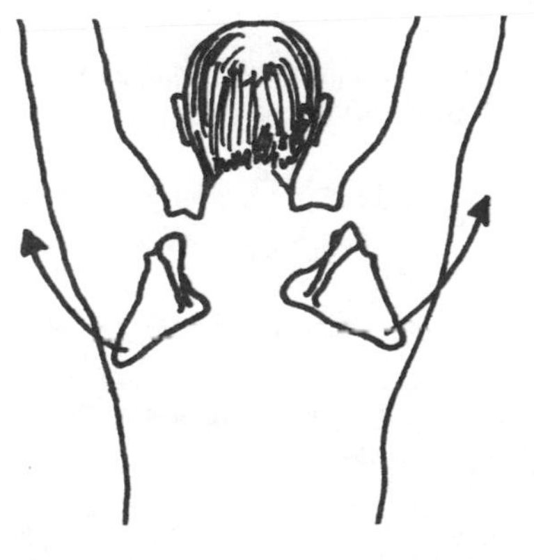

Backward rotation of the scapula

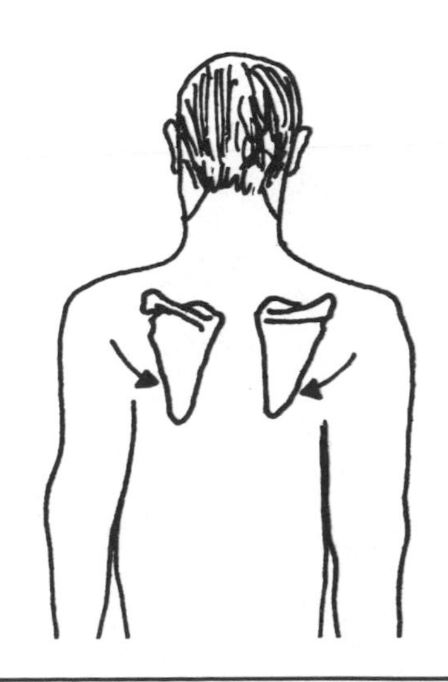

- gravity
- graded relaxation of trapezius (upper fibres) (p. 27) and serratus anterior (lower fibres) (p. 28)
- levator scapulae (p. 26)
- rhomboids (p. 29)
- pectoralis minor (p. 29)
- trapezius (lower fibres) (p. 27)

Study tasks

- Observe a colleague performing all scapular movements as shown.
- Highlight the names of the muscles producing the movements, and colour the position of the scapula in each case.
- Study the details of each muscle with reference to the page numbers given.

Levator scapulae ('muscle which lifts the scapula')

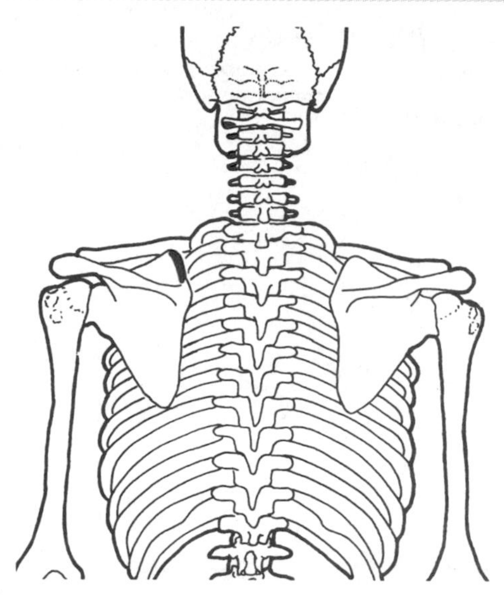

Attachments

- Posterior tubercles of TPs C1–4.
- Medial border of scapula between the superior angle and the spine.

Nerve supply

- C3, C4 and the dorsal scapular nerve (C5).

Functions

- Elevation of the scapula.
- Backward rotation of the scapula (from a position of forward rotation).
- Sidebending of the cervical spine, if the scapula is fixed.

Study tasks

- Shade lines between the attachment points and consider the muscle functions.
- Decide which attachment points act as the 'origin' and 'insertion' in each case (see p. 11).

Trapezius ('trapezium-shaped muscle')

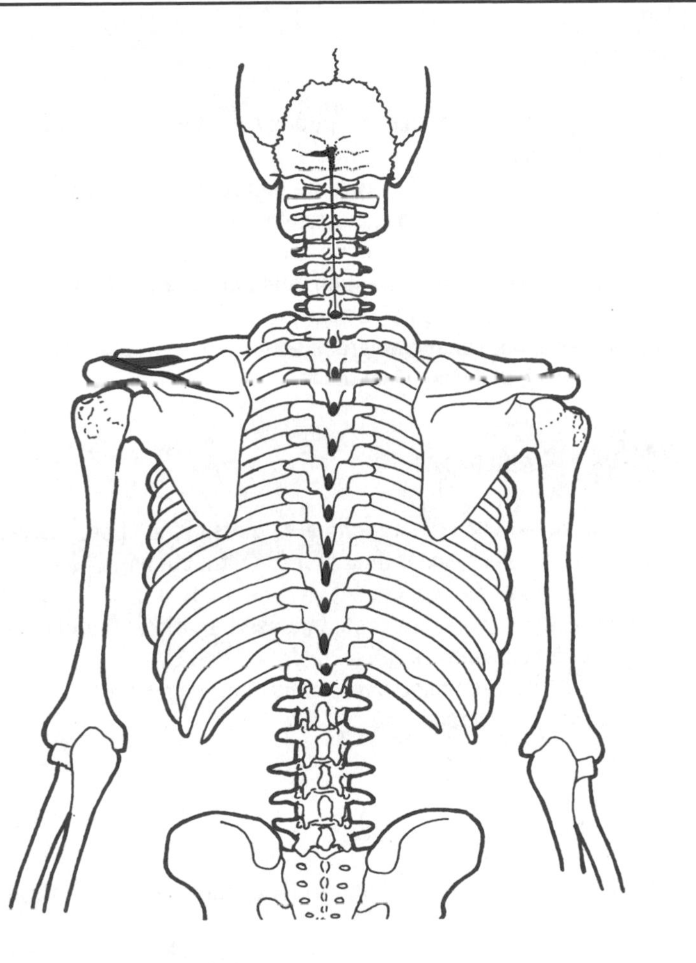

Attachments

Upper fibres:

- Medial part of superior nuchal line and external occipital protuberance.
- Ligamentum nuchae (the ligament connecting cervical SPs from the occiput to C7).

Middle fibres:

- SPs of C7–T6.

Lower fibres:

- SPs of T7–T12.
- All fibres converge onto the lateral part of the clavicle, the acromion process, and spine of the scapula.

Nerve supply

Cranial nerve XI (motor); C3, C4 (sensory).

Functions

- ♦ Unilaterally, the upper fibres produce elevation and forward rotation of the scapula.
- ♦ Unilaterally, with scapula fixed, the upper fibres produce sidebending of the cervical spine with slight rotation to the opposite side.
- ♦ The middle fibres produce retraction of the scapula.
- ♦ The lower fibres assist retraction, and backward rotation of the scapula (from a position of forward rotation).
- ♦ Bilaterally, if both scapulae are fixed, all fibres produce extension of the cervical and thoracic spine.

Study tasks

- Shade lines between the attachment points of the upper, middle and lower fibres, and consider their individual functions.
- Observe and test the muscle activity in the upper fibres on a colleague. Slight pressure should be used to restrain sidebending of the neck and elevation of the shoulder on one side.
- Decide which attachment point acts as origin and which as insertion in the case of each function.

Serratus anterior ('saw-toothed muscle at the front')

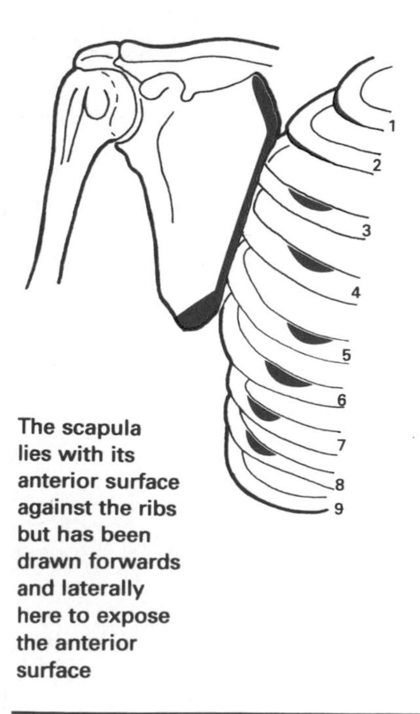

The scapula lies with its anterior surface against the ribs but has been drawn forwards and laterally here to expose the anterior surface

Attachments

- ■ Outer and upper borders of ribs 1–8 (occasionally also ribs 9 and 10) anteriorly.
- ■ Anterior surface of the medial border of the scapula.

Nerve supply

Long thoracic nerve (C5, C6, C7).

Functions

- ♦ Protraction of the scapula.
- ♦ Lower fibres produce depression and forward rotation of the scapula.
- ♦ Respiratory functions (p. 177).

Study tasks

- Shade lines between the attachment points and consider the muscle functions.
- Observe the muscle on a colleague with arms outstretched in front, as if in a diving pose.

Pectoralis minor ('smaller chest muscle')

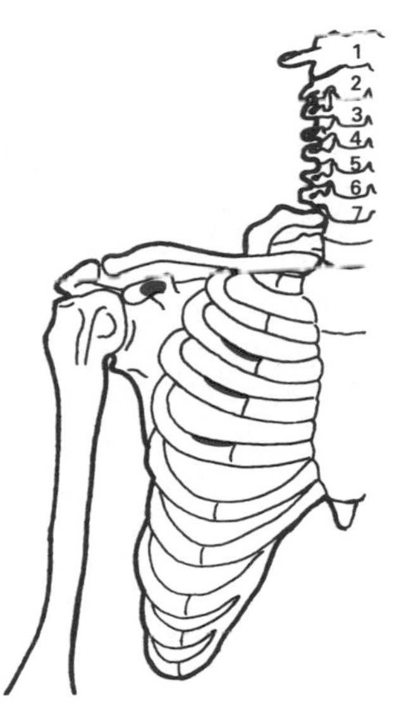

Attachments

- Ribs 3, 4 and 5 near their costal cartilages.
- Coracoid process of the scapula.

Nerve supply

Medial and lateral pectoral nerve (C6, C7, C8).

Functions

- Depression, protraction and backward rotation of the scapula from the front.
- Respiratory functions (p. 175).

Study tasks

- Shade lines between the attachment points and consider the muscle functions.
- Carefully palpate the coracoid process below the acromioclavicular joint on either yourself or a colleague.
- Palpate the active muscle just below the coracoid process by moving the shoulder forward and protracting the scapula.

Rhomboideus major and minor ('the rhomboids'; 'larger and smaller rhombus-shaped muscles')

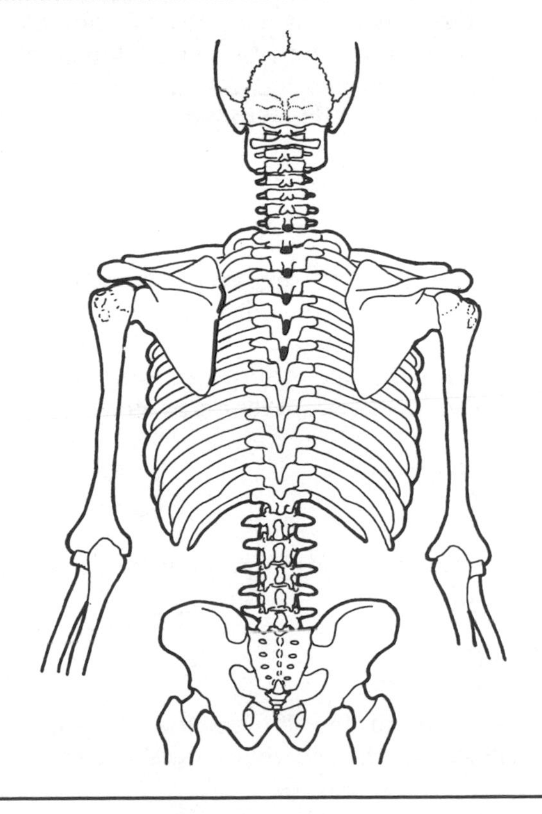

Attachments

- SPs of C7–T1 (minor).
- SPs of T2–5 (major).
- Medial border of scapula at root of its spine to inferior angle (minor lies superior to major).

Nerve supply

Dorsal scapular nerve (C4, C5).

Function

- Retraction with slight elevation of scapula.

Study tasks

- Shade lines between the attachment points and consider the muscle functions.
- Observe the rhomboids in action on a colleague, strongly retracting their shoulders.

Movements of the entire shoulder girdle

Shoulder movements are dominated by the action of the glenohumeral ball and socket joint, which allows flexion and extension through a transverse axis; abduction and adduction through an antero-posterior axis; medial and lateral rotation through a longitudinal axis; and the combined movement of circumduction.

Flexion

The full range is 180°, but requires forward rotation of the scapula and swivel along the shaft of the clavicle as movement takes place. The early stages of movement are produced by muscles acting on the glenohumeral joint, whilst the later stages are assisted by the scapular muscles which have already been described.

Phase 1: 0°–60°

This phase of movement is dominated by movement at the glenohumeral joint only.

Phase 2: 60°–120°

In order to achieve further flexion the scapula moves into forward rotation which allows the glenoid surface of the scapula to face upwards. This is accompanied by rotation along the shaft of the clavicle and can be palpated as movements at the sternoclavicular and acromioclavicular joints. The glenohumeral joint is often regarded as contributing two-thirds, and the scapula one-third, of total shoulder movement. The muscles which produce this movement are those in Phase 1, plus those which forwardly rotate the scapula.

Phase 3: 120°–180°

The final phase of movement is achieved by all of the flexor muscles with additional slight sidebending of the spinal column by the erector spinae muscles of the contralateral side.

Flexion of the shoulder

Phase 3: 120°– 180°

Phase 1 and 2 muscles, plus muscles which produce vertebral sidebending (contralateral side) (p. 212)

Phase 2: 60°– 120°

Phase 1 muscles, plus muscles which forwardly rotate the scapula:
- trapezius (upper fibres) (p. 27)
- serratus anterior (lower fibres) (p. 28)

Phase 1: 0°– 60°
- pectoralis major (clavicular part) (pp. 32–33)
- anterior deltoid (p. 32)
- coracobrachialis (pp. 33–34) assisted by:
- biceps brachii (p. 34)

Final phase of flexion (or abduction) of the shoulder girdle: 120°– 180°)

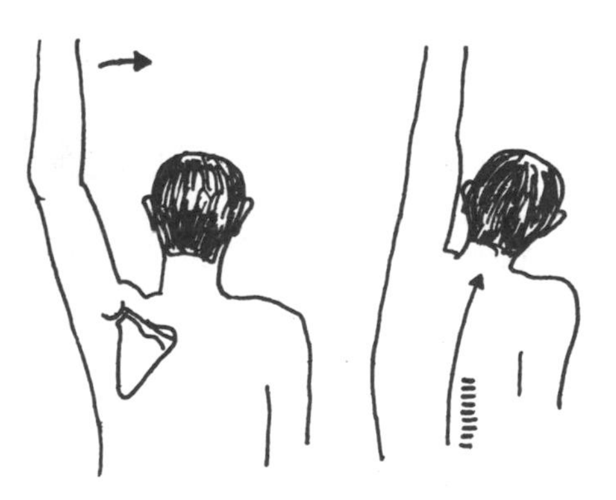

Study tasks

- Highlight the names of the muscles shown.
- Study the details of each muscle with reference to the page numbers given.

Deltoid ('delta- or triangular-shaped muscle')

This convergent muscle is Nature's 'shoulder pad', and is functionally subdivided into three parts. Note that only the anterior fibres are relevant to shoulder flexion, but full details of the muscle are included here for further reference.

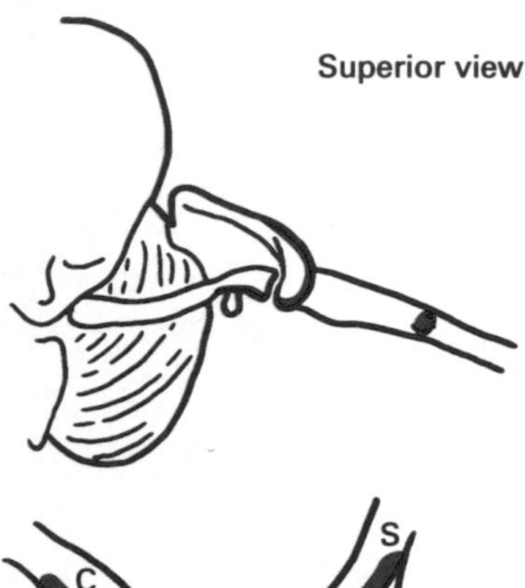

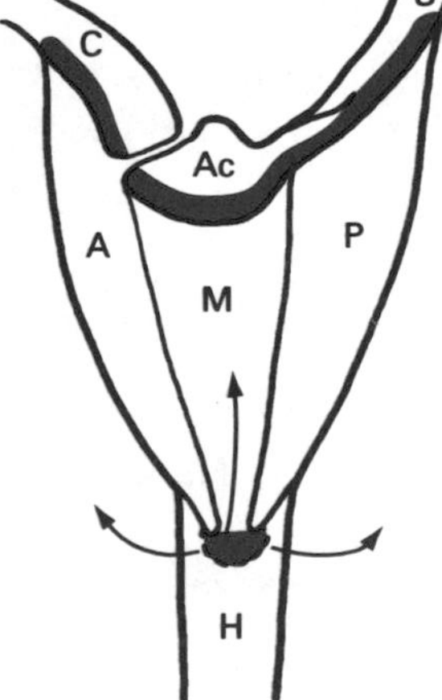

The three parts of deltoid muscle

- A Anterior fibres
- M Middle fibres (multipennate)
- P Posterior fibres
- S Spine of scapula
- Ac Acromion
- C Clavicle
- H Humerus
- ‿ Directions of shoulder movement

Attachments

- Lateral part of the clavicle.
- Lateral border of the acromion.
- Crest of the spine of the scapula.
- Deltoid tuberosity of the humerus.

Nerve supply

Axillary nerve (C5, C6).

Functions

- The middle fibres and whole muscle abduct the shoulder.
- The anterior fibres produce flexion and medial rotation of the shoulder.
- The posterior fibres produce extension and lateral rotation of the shoulder.

Study tasks

- Shade lines between the attachment points and consider the muscle functions.
- Test the anterior fibres on a colleague's arm abducted to 90° with elbow flexed and thumb pointing down into medial rotation. Resist just above the elbow while they push forward into flexion.
- The middle fibres may be tested as above, in 90° abduction, resisting while they push upwards into further abduction. No rotation is necessary.
- The posterior fibres may be tested as above but with thumb pointing upwards into lateral rotation. Resist while they push back into extension.
- Consider which class of lever is represented by the action of the middle fibres in pure abduction (p. 18).
- Why are only the middle fibres apparently multipennate?

Pectoralis major ('large chest muscle')

Colloquially called 'pecs' in gymnasia and health clubs, this muscle gives definition to the upper chest, and functionally consists of two parts: **clavicular** and **sternocostal**.

Attachments

- Medial part of the clavicle (clavicular part).
- Costal cartilages 1–6 and sternum (sternocostal part).
- Lateral lip of the bicipital groove of the humerus.

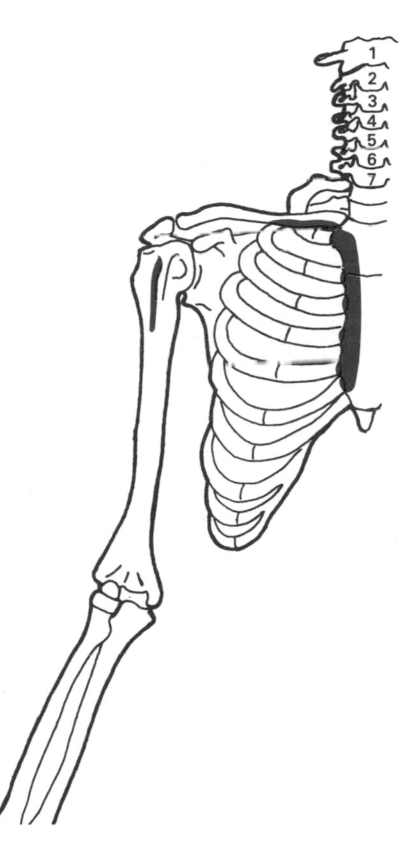

Nerve Supply

Lateral and medial pectoral nerves (C5, C6, C7, C8, T1).

Functions

- The clavicular part flexes the shoulder from the anatomical position.
- The sternocostal part flexes the shoulder if the humerus is fully extended.
- Both parts produce adduction and medial rotation of the shoulder.
- Respiratory functions (p. 174).

Study tasks

- Shade lines between the attachment points and consider the muscle functions. (See *Gray's Anatomy* (1995) for precise details about muscle arrangement).
- Test the clavicular part on a colleague with their arms outstretched and palms turned out so that the shoulder is in medial rotation. Gently force their arms apart as they resist, and observe the muscle contract at the clavicle.
- Test the sternocostal part on yourself in front of a mirror by placing hands on hips and pushing the shoulders into adduction. The sternocostal part can be seen to contract.

Coracobrachialis ('muscle of the arm and coracoid process')

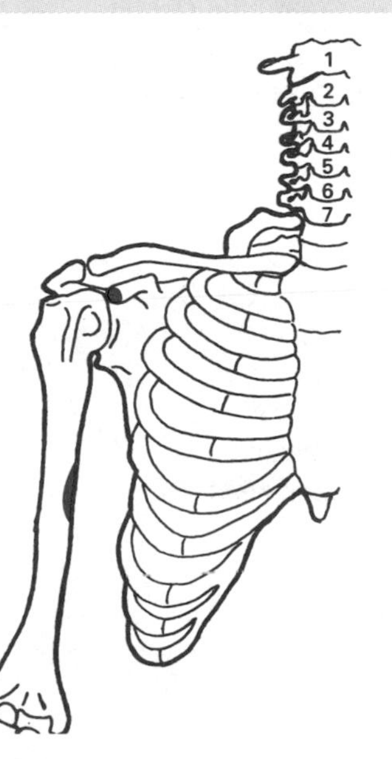

Attachments

- Tip of the coracoid process.
- Medial border of the mid-shaft of the humerus.

Nerve supply

Musculocutaneous nerve (C5, C6, C7).

Functions

- Flexion of the shoulder.
- Adduction of the abducted shoulder.

Study tasks

- Shade lines between the attachment points and consider the muscle functions.
- The muscle can be tested, although it is not visible, by asking a colleague to flex the shoulder with slight abduction and lateral rotation. Resist while they try to push their arm back to the anatomical position.

Biceps brachii ('two-headed muscle of the arm')

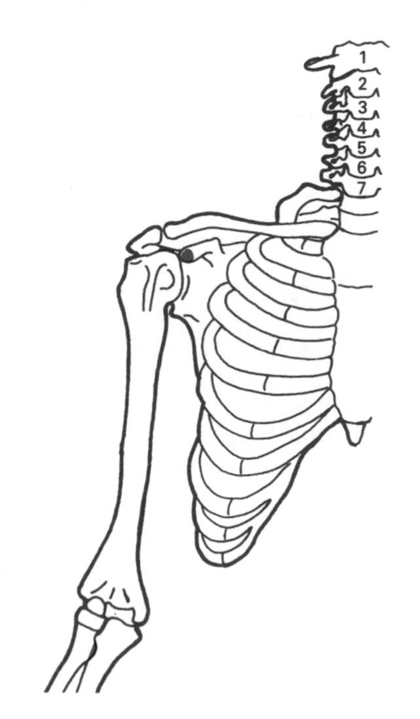

Note: Biceps is only a limited flexor of the shoulder. Its main functions are supination and flexion of the elbow (p. 49).

Attachments

- Long head: from the supraglenoid tubercle of the scapula.
- Short head: from the coracoid process of the scapula.
- The radial tuberosity *and* the bicipital aponeurosis in the deep fascia of the forearm.

Nerve supply

Musculocutaneous nerve (C5, C6).

Functions

- Assists flexion of the shoulder.
- Supination of the forearm and hand.
- Flexion of the elbow.

Study task

- Shade lines between the attachment points, but only consider its shoulder functions here.

Extension

Active range is normally about 50° and can be divided into two phases. Phase 1 (approximately 0°–20°) occurs at the glenohumeral joint, but from 20° to 50° (Phase 2) there is accompanying retraction and backward rotation of the scapula.

Extension of the shoulder

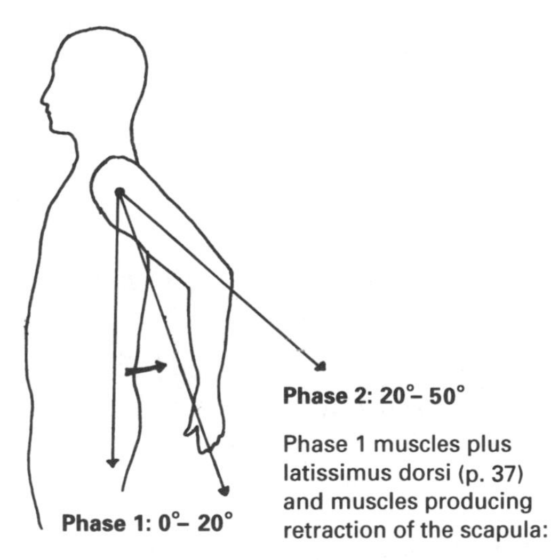

Phase 2: 20°– 50°

Phase 1 muscles plus latissimus dorsi (p. 37) and muscles producing retraction of the scapula:

- posterior deltoid (p. 32)
- teres major (p. 36)

- trapezius (middle fibres) (p. 27)
- rhomboids (p. 29)

Note:
If extension commences from a position of full flexion, the following muscles are said to be active:

- pectoralis major (sternocostal part) (pp. 32–33)
- latissimus dorsi (p. 37)

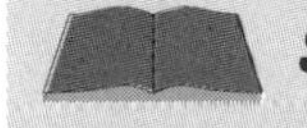

Study tasks

- Highlight the names of the muscles shown.
- Study the details of each muscle with reference to the page numbers given.

Teres major ('larger smooth muscle')

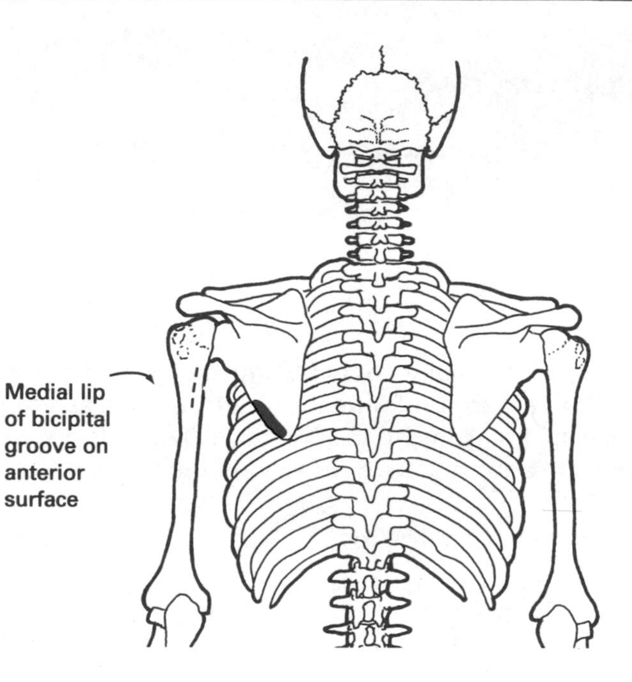

Attachments

- Lower part of lateral border of the scapula.
- Medial lip of bicipital groove (biceps tendon groove) of the humerus.

Nerve supply

Lower subscapular nerve (C6, C7).

Functions

- Extension of the shoulder.
- Medial rotation of the shoulder.

Study tasks

- Shade lines between the attachment points, and consider the muscle functions.
- Observe and test on a muscular colleague by asking them to place their hands on their hips and extend their shoulders while you stand behind and restrain their elbows. The muscle should be visible along the lower lateral border of the scapula.

Latissimus dorsi ('widest muscle of the back')

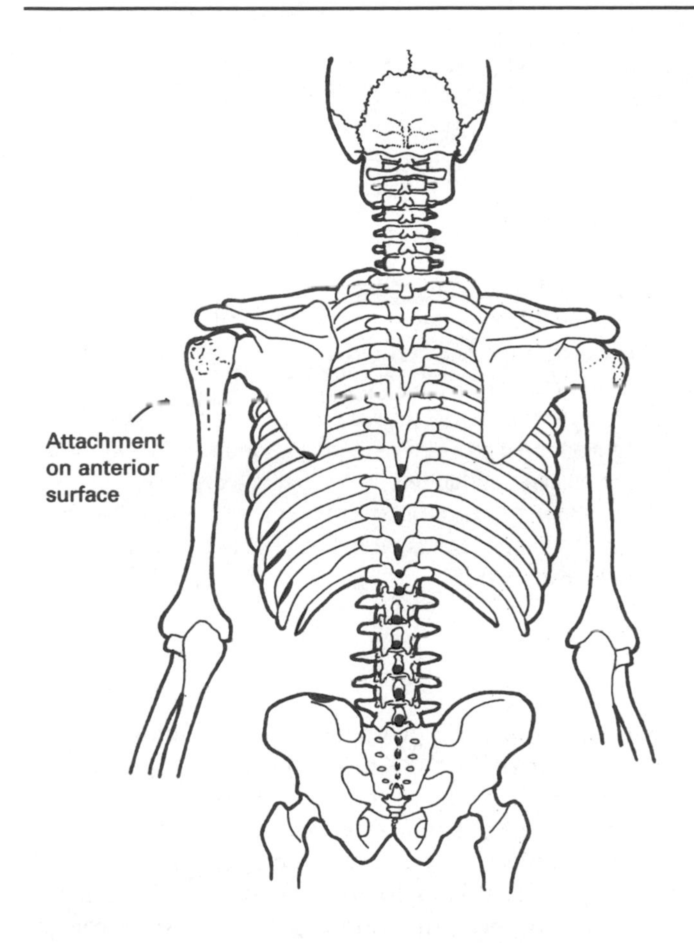

This is the largest, widest muscle of the back, which inserts into the humerus and therefore becomes a powerful extensor, adductor and medial rotator muscle of the shoulder. It is often colloquially referred to as 'the lats' in its plural form.

Attachments

- External iliac crest and the thoracolumbar fascia.
- SPs of T7–S5.
- Lower four ribs (tendinous slips).
- Inferior angle of the scapula.
- Floor of the bicipital groove (biceps tendon groove of the humerus).

Nerve supply

Thoracodorsal nerve (C6, C7, C8).

Functions

- ♦ Powerful extension of the shoulder especially from a position of flexion (as in rowing).

- Medial rotation of the shoulder.
- Adduction of the shoulder.
- Respiratory functions (pp. 176 and 182).

Study tasks

- Shade lines from the lower attachment points to converge on the bicipital groove. The muscle attaches anteriorly, and therefore twists as it approaches the bicipital insertion point. (See *Gray's Anatomy* (1995) for precise details.)
- Test the muscle on a colleague's arm abducted to approximately 30° with the thumb pointing down to give medial rotation. Hold their arm while they try to pull into adduction.
- Compare the shoulder and respiratory functions of this muscle, and consider which attachment points form the origin and which the insertion in each case.

Abduction

Active abduction has a total range of 180°, and like flexion can be divided into three phases. Phase 1 (0°–90°) is usually performed with the palm of the hand facing downwards. This means that the greater tuberosity of the humerus comes into contact with the glenoid labrum and coracoacromial ligament at about 90°, and further abduction is difficult. In Phase 2 (90°–150°) the arm is laterally rotated to 'unlock' the humerus (the lateral rotator muscles are described separately). Abduction proceeds with the help of forward rotation of the scapula and swivel of the clavicle. The glenohumeral joint contributes 2° for every 3° of movement, as in flexion. In Phase 3 (150°–180°) the same position has been reached as in final flexion. Slight sidebending of the vertebral column is necessary to reach 180°. If both arms are fully abducted, extension of the lumbar spine is necessary.

Abduction of the shoulder

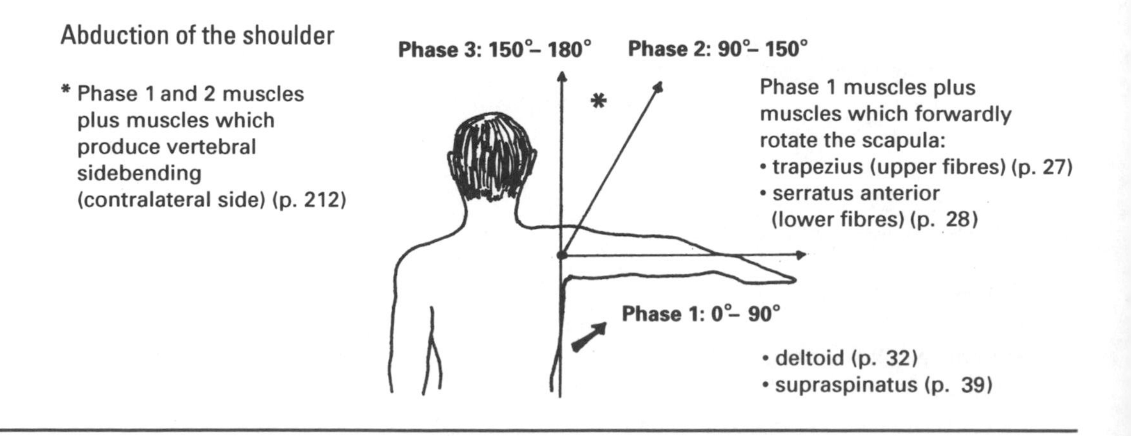

Note: The rotator cuff muscles probably play an important role in abduction by fixing and stabilising the head of the humerus, while deltoid exerts leverage on the shaft.

Study tasks

- Highlight the names of the muscles shown.
- Study the details of each muscle with reference to the page numbers given.

Supraspinatus ('muscle above the spine of the scapula')

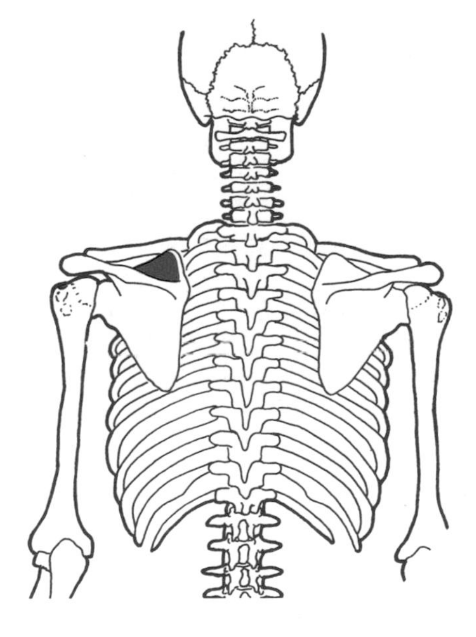

Attachments

- Supraspinous fossa on the scapula.
- Superior facet on the greater tuberosity of the humerus.

Nerve supply

Suprascapular nerve (C4, C5, C6).

Functions

- Once regarded as the main initiator of abduction, the muscle is now regarded as assisting deltoid in abduction.
- Part of the rotator cuff mechanism (pp. 23–24).

The subacromial surface

Subacromial surface of the scapula

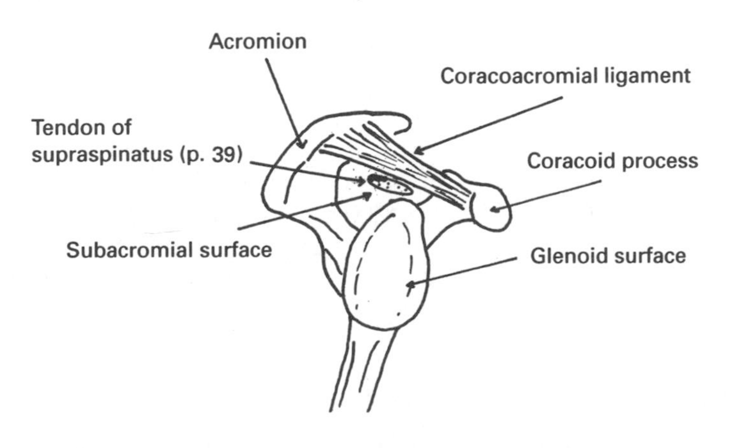

The tendon of supraspinatus slides below the coracoacromial arch formed by the coracoacromial ligament, with the protective subacromial bursa lying between the tendon and the acromion process. Injury or degeneration in these structures causes inflammatory swelling in a confined space, with pain in the first phase of abduction, as the greater tuberosity of the humerus approaches the subacromial surface.

> ***Study tasks***
>
> - Shade lines between the attachment points of supraspinatus and consider its functions.
> - Shade the diagram of the subacromial surface and highlight the features shown.

Adduction of the shoulder (posterior view)

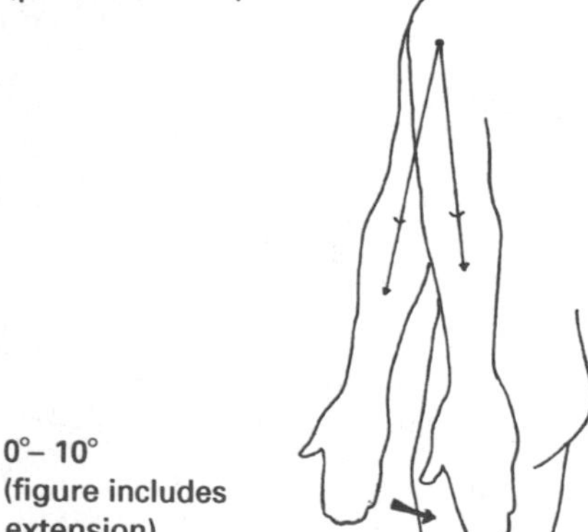

- latissimus dorsi (p. 37)
- pectoralis major (pp. 32–33)
- teres major (p. 36)
- coracobrachialis (pp. 33–34)

Adduction

Pure adduction is not possible because the torso obstructs arm movement in the anatomical position. Adduction must therefore be combined with either flexion or extension, but often refers to 'relative' adduction which is a return to the anatomical position from any point of abduction.

Active adduction in flexion reaches up to 45° and in extension 10°. The movement is aided by gravity in the case of relative adduction.

> ***Study tasks***
>
> - Highlight the names of the muscles shown.
> - Study the details of each muscle with reference to the page numbers given.

Medial rotation of the shoulder (0°–100° if hand is taken posteriorly)

- pectoralis major (pp. 32–33)
- anterior deltoid (p. 32)
- latissimus dorsi (p. 37)
- teres major (p. 36)
- subscapularis (p. 41)

Medial rotation

Rotation of the humerus through the longitudinal axis of the shaft produces medial rotation and the hand is carried towards the midline. In practice the torso often obstructs movement, but if the hand is carried behind the back (the 'half nelson' position), a range of 100° may be possible. This position is uncomfortable due to the pull on the lateral rotator tendons (see below).

> ***Study tasks***
>
> - Highlight the names of the muscles shown.
> - Study the details of each muscle with reference to the page numbers given.

Subscapularis ('muscle under the scapula')

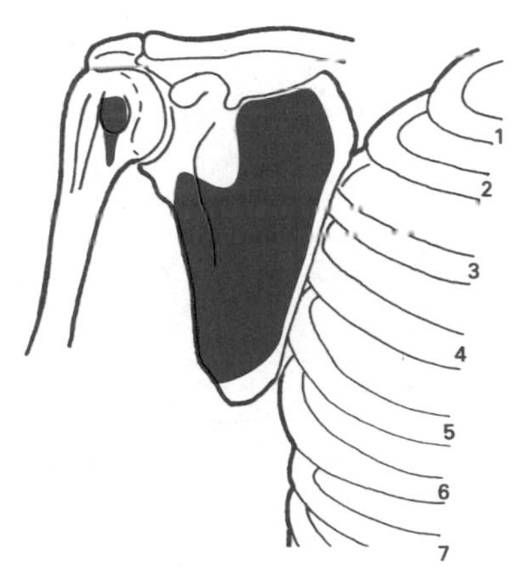

The scapula lies with its anterior surface against the ribs but has been drawn forwards and laterally here to expose the anterior surface

Attachments

- Anterior surface of the scapula.
- Lesser tuberosity of the humerus (sole occupant).

Nerve supply

Upper and lower subscapular nerves (C5, C6, C7).

Functions

- Medial rotation of the shoulder when the arm is by the side.
- Part of the rotator cuff mechanism (pp. 23–24).

Study tasks

- Shade lines between the attachment points and consider the muscle functions.
- It is not easy to isolate individual rotator muscles, but aim to test subscapularis on a colleague's arm held at their side with elbow flexed. They should try to push their wrist towards the midline while you resist.
- Subscapularis is an example of a multipennate muscle. Why do you think this should be so?

Lateral rotation

This movement through the longitudinal axis of the humerus carries the hand away from the midline, and has an active range up to about 80° which can be increased by forward rotation of the scapula when brushing the hair, or even by scapular retraction if the arm is by the side.

Lateral rotation of the shoulder (0°– 80°)

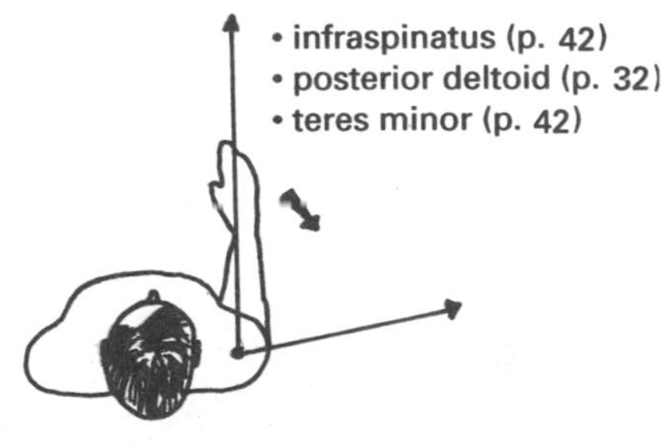

Study tasks

- Highlight the names of the muscles shown.
- Study the details of the muscles with reference to the page numbers given.

Infraspinatus ('muscle below the spine of the scapula')

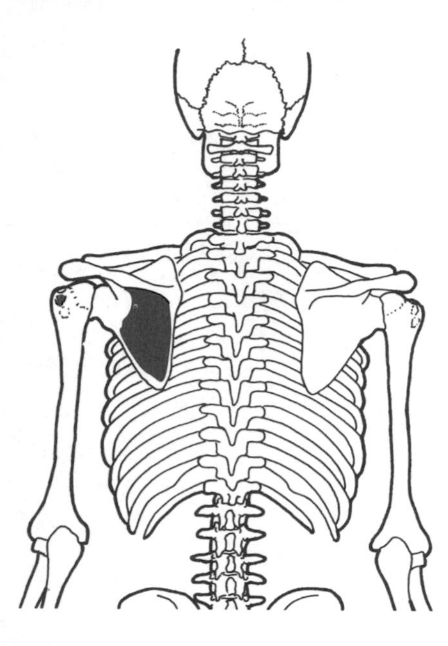

Attachments

- Infraspinous fossa of the scapula.
- Middle facet on the greater tuberosity of the humerus.

Nerve supply

Suprascapular nerve (C4, C5, C6).

Functions

- Lateral rotation of the shoulder.
- Part of the rotator cuff mechanism (pp. 23–24).

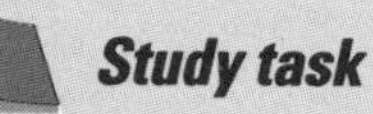

Study task

- Shade lines between the attachment points and consider the muscle functions.

Teres minor ('smaller smooth muscle')

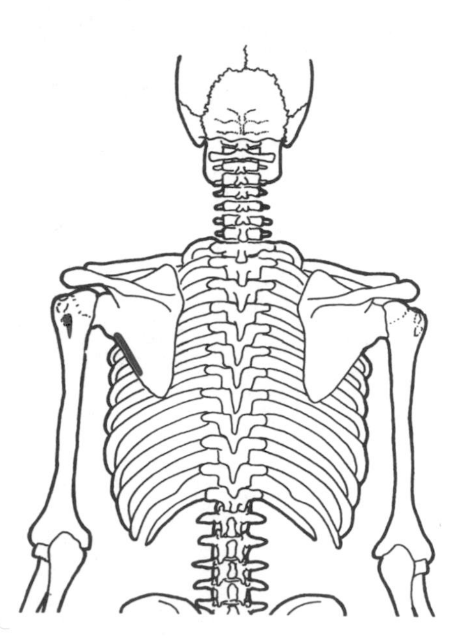

Attachments

- Upper part of lateral border of the scapula.
- Lowest facet on the greater tuberosity of the humerus.

Nerve Supply

Axillary nerve (C4, C5, C6).

Functions

- Lateral rotation of the shoulder.
- Part of the rotator cuff mechanism (pp. 23–24).

Study tasks

- Shade lines between the attachment points and consider the muscle functions.
- Test lateral rotation of the shoulder on a colleague with arm abducted to 90°, elbow flexed, so that you resist movement as they push their wrist posteriorly. Stand behind and steady their movement. Palpate the scapula for activity in teres minor and infraspinatus.

Circumduction

This is a movement in which the arm describes the base of a cone, with the glenohumeral joint forming the apex. It combines all movements described above, and increases in scope with added scapular movement.

Circumduction of the shoulder

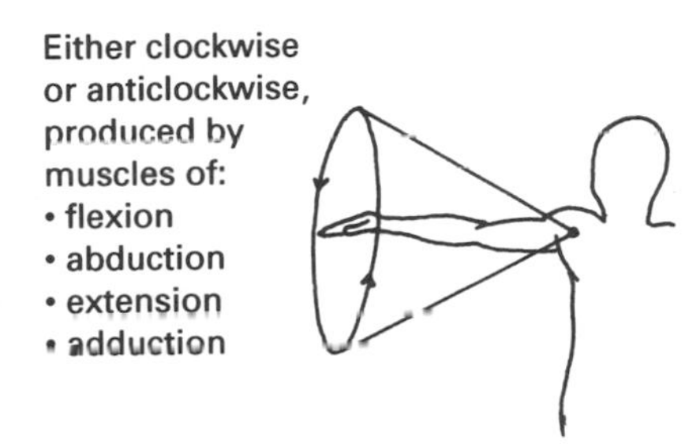

Study tasks

- Highlight the names of the individual movements producing circumduction.
- Revise all shoulder movements by standing behind a colleague and asking them to perform:
  pure scapular movements;
  full shoulder girdle movements including circumduction, and observe how the scapula moves.
- Palpate rotation along the clavicle, and movements at the sternoclavicular and acromioclavicular joints as the shoulder moves.
- Listen with a stethoscope to the sounds that these joints produce on a healthy colleague (most important). You may be surprised by what you hear. Try to interpret these sounds with reference to the palpation task just performed, and also with reference to the details of the shoulder joints given on p. 22.

In all movements of the shoulder, the clavicle is stabilized by the action of subclavius muscle. It lies deep in the clavipectoral fascia, and cannot be palpated.

Subclavius ('muscle below the clavicle')

Attachments

- First rib and costal cartilage.
- Inferior surface of middle clavicle.

Nerve supply

Nerve to subclavius (C5, C6).

Function

- Stabilizes the clavicle during movements of the shoulder.

(Inferior clavicle shown)

Study task

- Shade lines between the attachment points and consider the muscle function.

Chapter 5

The elbow

Introduction

The elbow is a compound synovial joint consisting of articulations between the distal end of the humerus and the proximal ends of the radius and ulna.

The bones and joints

The main joint consists of a synovial hinge arrangement between the trochlea of the humerus and the trochlear notch of the ulna (the humero-ulnar joint). There is also a separate articulation between the capitulum of the humerus and the head of the radius (the humeroradial joint).

Bones of the elbow

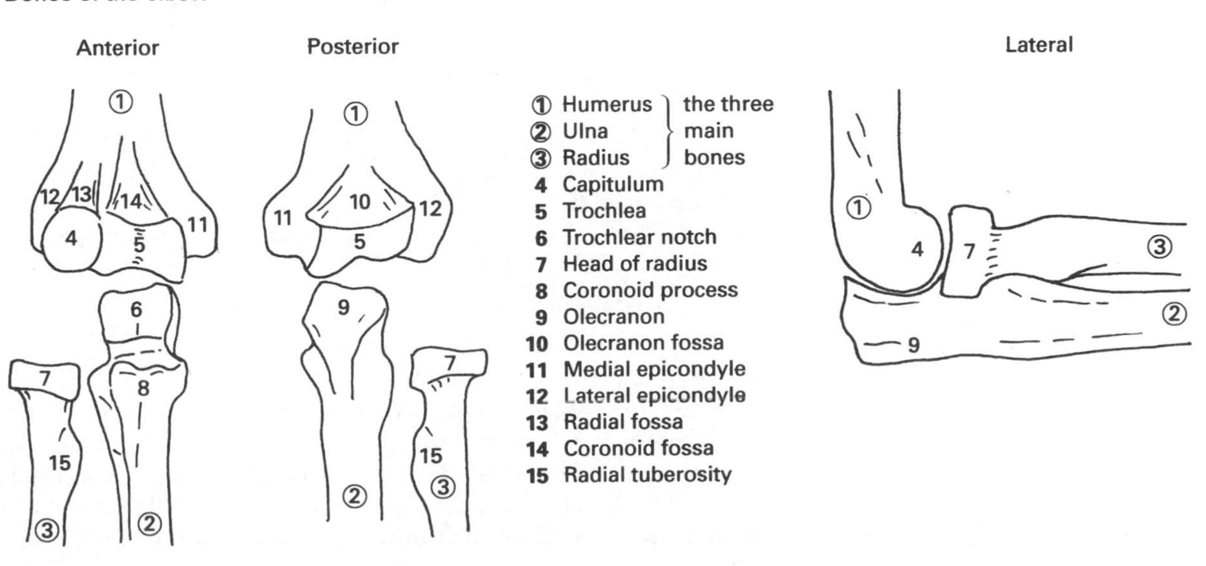

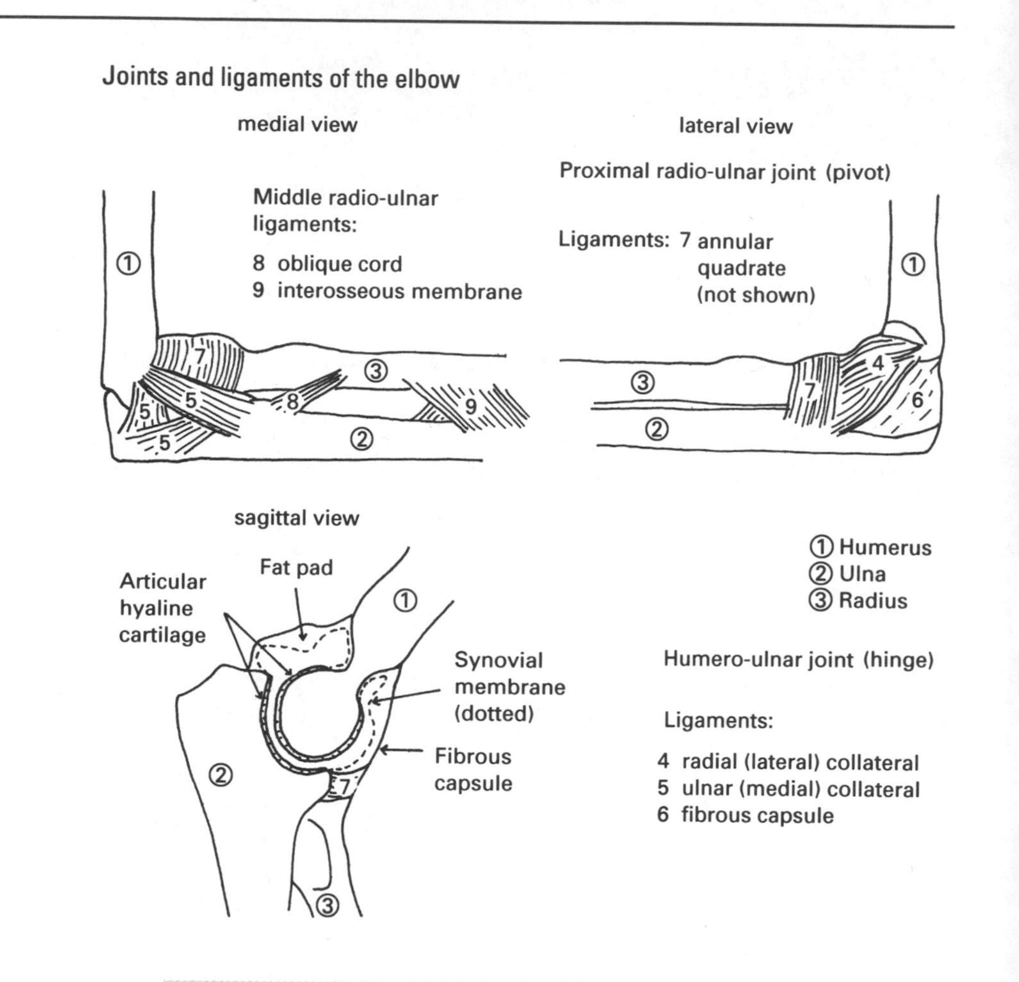

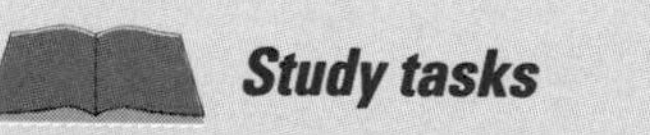

Study tasks

- Highlight the names of the bone features as indicated.
- Shade the ligaments and highlight the details of the joints.
- Obtain bone specimens and identify the features shown.
- Palpate flexion and extension on a colleague's elbow. In particular, identify the bony 'end feel' of full extension in the straight arm, as the olecranon of the ulna finds its resting place in the olecranon fossa of the humerus. Perform this movement with care: it is painful if done forcefully or rapidly.

The rationale of this arrangement is to provide strong flexion and extension through a transverse axis, but the human elbow is capable of much more than this. There is also a pivot joint between the circular head of the radius and a reciprocal notch on the ulna (proximal radio-ulnar joint), which allows for the rotatory movements of pronation and supination of the hand.

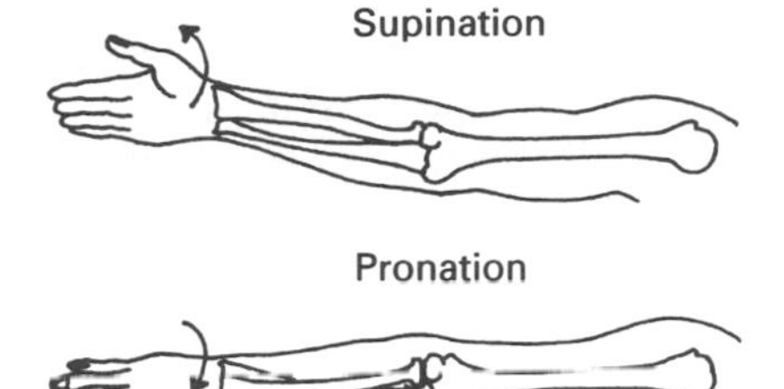

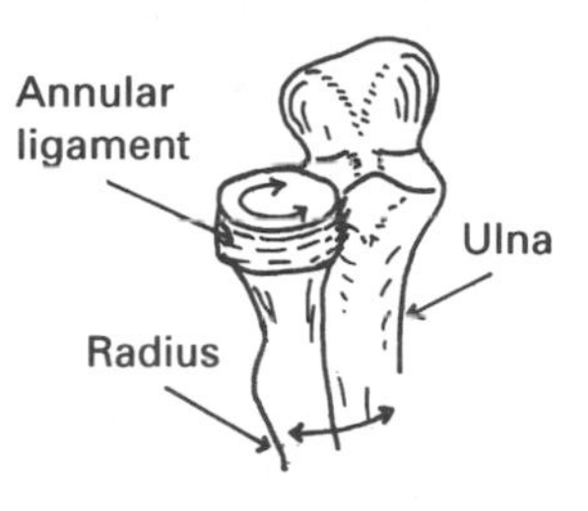

The proximal radio-ulnar joint

(further details pp. 53–57)

Study tasks

- Shade the two bones of the pivot joint in separate colours, and the annular ligament in another colour.
- Palpate the circular head of the radius in a colleague's elbow (it is enclosed by the surrounding annular ligament), and then palpate the movements of supination and pronation at this point.

The Humero-ulnar and humeroradial joints

The trochlear notch on the ulna does not fit perfectly with the trochlea of the humerus since the medial lip or flange of the trochlea is wider than its lateral lip. Nor is the coronoid process of the ulna at right angles with the shaft of the ulna. Thus, in extension, the ulna deviates laterally, giving the angulation at the elbow joint known as 'the carrying angle', when the hand is supinated.

This angle between the shafts of the humerus and ulna varies from about 160° to 175°, with the greater angulation usually seen in the female. This is probably due to the greater width of the pelvis.

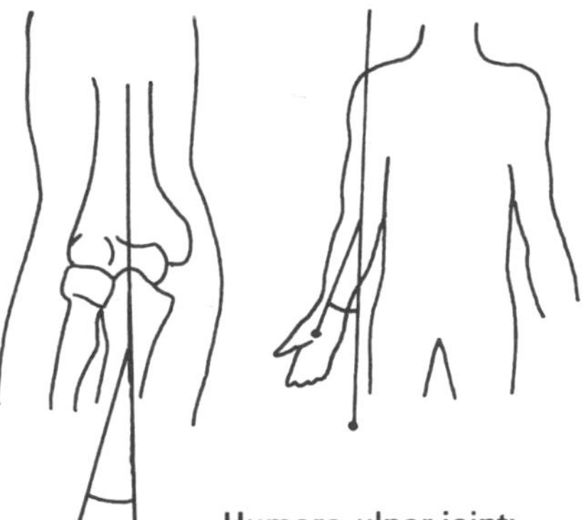

Humero-ulnar joint: 'the carrying angle'

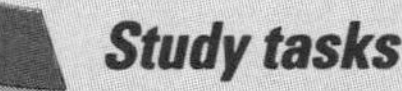

Study tasks

- Shade the carrying angle in colour.
- Examine the variation in the carrying angle between male and female colleagues.

Movements of the main humero-ulnar joint

The main hinge mechanism of the humerus and ulna allows only the movements of flexion and extension through the transverse axis.

However, as already demonstrated, the radius is able to rotate across the ulna pivoting at the proximal and distal radio-ulnar joints in the movements known as pronation and supination (pp. 53–57).

Flexion

Flexion of the elbow

(0°– 145°)
- brachialis (p. 48)
- biceps brachii (p. 49)
- brachioradialis (p. 50)

assisted by:
- pronator teres (p. 57)

Study tasks

- Highlight the names of the muscles shown.
- Study the details of each muscle with reference to the page numbers given.

Brachialis ('muscle of the arm')

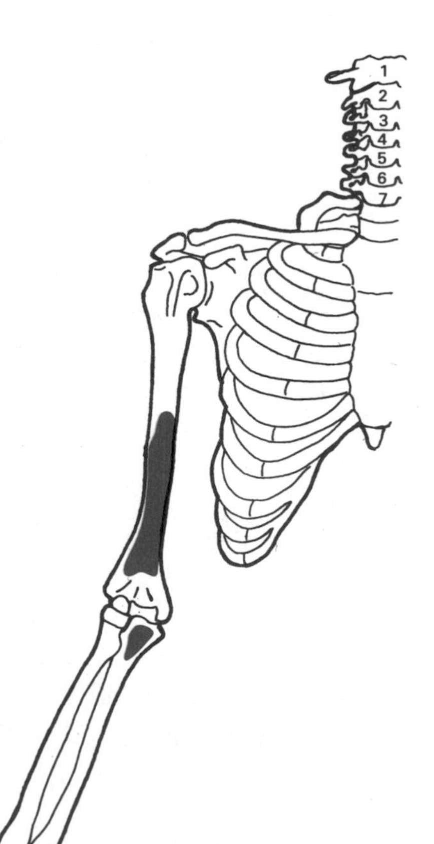

Attachments

- ■ Lower part of anterior surface of the humerus.
- ■ Below the coronoid process of the ulna.

Nerve supply

Musculocutaneous nerve (C5, C6): the lateral portion is partly supplied by the radial nerve (C7).

Function

- ♦ Flexion of the elbow joint.

Study tasks

- Shade lines between the attachment points and consider the muscle function.
- Which class of lever is this (p. 18)?

Biceps brachii ('two-headed muscle of the arm')

Biceps is a powerful muscle of both flexion and supination of the elbow. It takes its name from its two attachment points or 'heads'.

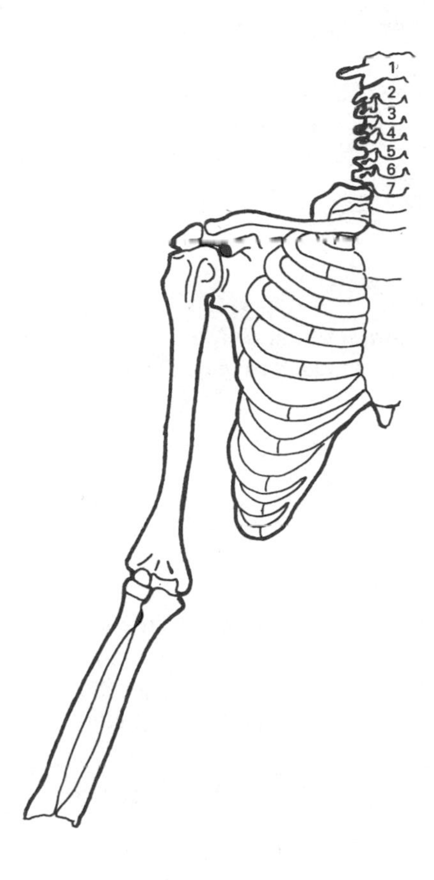

Attachments

- Long head: from the supraglenoid tubercle of the scapula.
- Short head: from the coracoid process of the scapula.
- The radial tuberosity **and** the bicipital aponeurosis in the deep fascia of the forearm. **Note:** this can be felt as a sharp edge of tendon on the medial side of a flexed elbow just below the belly of the muscle: the hand should be supinated.

Nerve supply

Musculocutaneous nerve (C5, C6).

Functions

- Flexion of the elbow.
- Supination of the forearm and hand.
- Assists flexion of the shoulder.

Study tasks

- Shade lines between the attachment points and consider the muscle functions.
- Test the flexor action of biceps and brachialis on a colleague with elbow flexed to 90°, and palm facing upwards forming a loose fist. Resist at the wrist as they attempt to flex their elbow.

Brachioradialis ('muscle of the arm and radius')

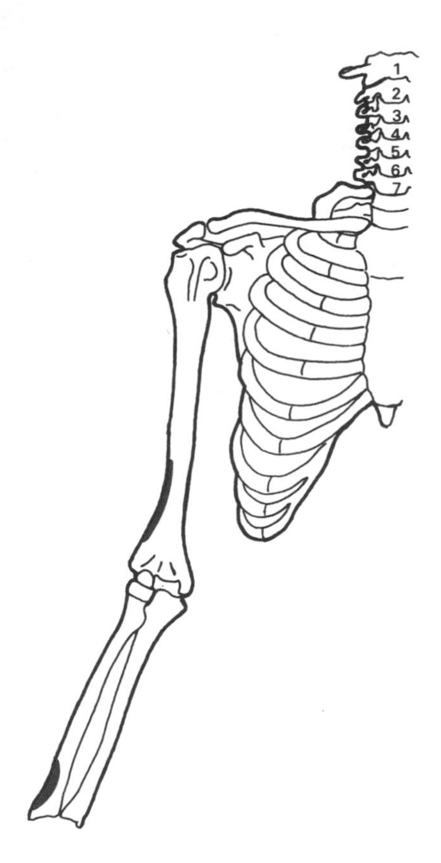

Attachments

- Upper part of the lateral supracondylar ridge of the humerus.
- Lateral side of radius above the styloid process.

Nerve supply

Radial nerve (C5, C6, C7).

Functions

- Flexion of the elbow particularly from the position midway between pronation and supination (as in a karate 'chop').
- Assists pronation from a position of full supination.
- Assists supination from a position of full pronation.

Study tasks

- Shade lines between the attachment points and consider the muscle functions.
- Identify which point acts as origin and which acts as insertion.
- Place your forearm in a position of full pronation and then supination, and consider why the muscle can assist both supination and pronation.
- Test the flexion action by placing a colleague's flexed forearm in the midway position described above, and resist flexion. The muscle contraction should be easily seen.

Extension

This brings the supinated hand away from the shoulder, and straightens the elbow. In the anatomical position full active extension has been reached at 0°, so the movement is essentially relative from some degree of flexion. A few extra degrees of 'hyperextension' may be noted, with up to 10° if ligamentous laxity is present.

Active extension usually involves gravity, plus the gradual relaxation of the flexor muscles described above. However, forcible extension of the elbow actively recruits the extensor muscles in activities such as throwing.

Study tasks

- Highlight the names of the muscles shown.
- Study the details of each muscle with reference to the page numbers given.

Extension of the elbow

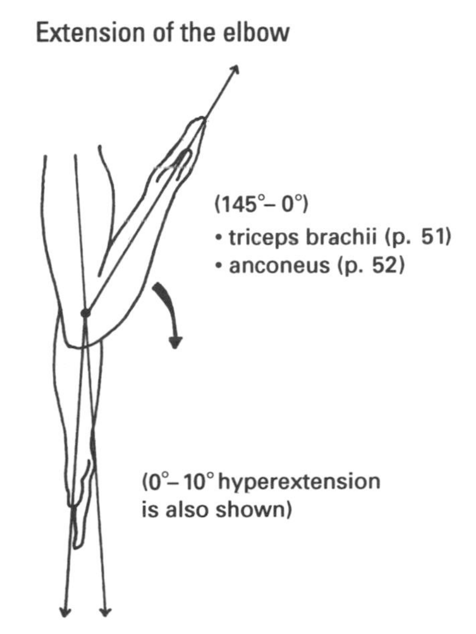

Triceps brachii ('three-headed muscle of the arm')

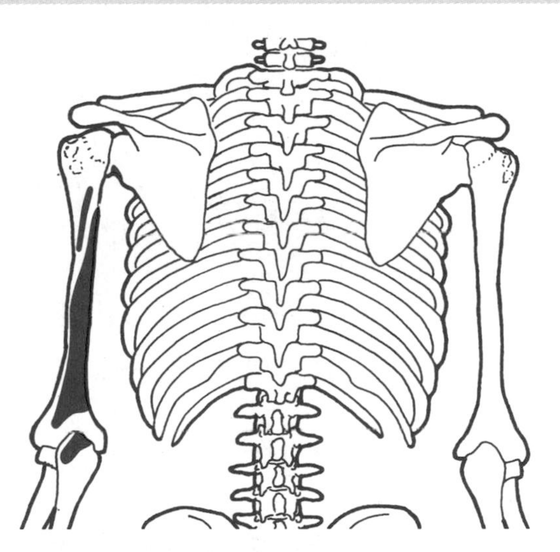

As the name implies, this muscle has three points of origin which then converge to join a common tendon which inserts into the posterior olecranon process of the ulna.

Attachments

- Long head: from the infraglenoid tubercle of the scapula.
- Lateral head: from the posterior part of the humerus, above the groove for the radial nerve.
- Medial head: from the posterior part of the humerus below the groove for the radial nerve.
- The posterior part of the olecranon process of the ulna.

Nerve supply

Radial nerve (C6, C7, C8).

Function

- Extension of the elbow.

Study tasks

- Shade lines between the attachment points and consider the muscle function.
- Test the muscle by flexing a colleague's elbow to about 45° and in slight abduction. Support triceps with one hand while you resist extension with the other.
- Palpate your own arm by placing your fingers on biceps and your thumb on triceps. Compare the activity of both triceps and biceps in the following situations:
  - slightly pushing up as if in a 'press-ups' position;
  - lowering a glass to a table;
  - throwing an object.

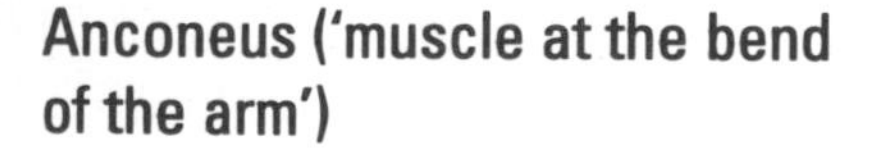

Anconeus ('muscle at the bend of the arm')

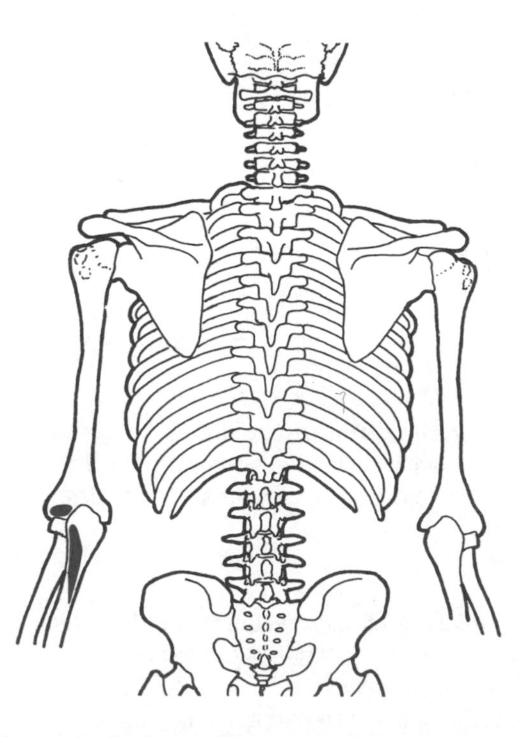

Anconeus is a relatively weak extensor: the main points to consider have been dealt with under 'triceps'.

Attachments

- Lateral epicondyle of the humerus, and capsule of the elbow joint.
- Lateral aspect of the olecranon process and posterior surface of the ulna.

Nerve supply

Radial nerve (C7, C8, T1).

Function

- Extension of the elbow.

Study task

- Shade lines between the attachment points and consider the muscle function.

Movements at the radio-ulnar joints (pronation and supination)

The shafts of the radius and ulna bones face each other below the elbow joined by a tough interosseous membrane which increases the stability of the forearm and acts as an attachment for muscles.

The radius is able to rotate across the ulna through an oblique longitudinal axis which pivots at the proximal and distal radio-ulnar joints allowing the movements of pronation and supination.

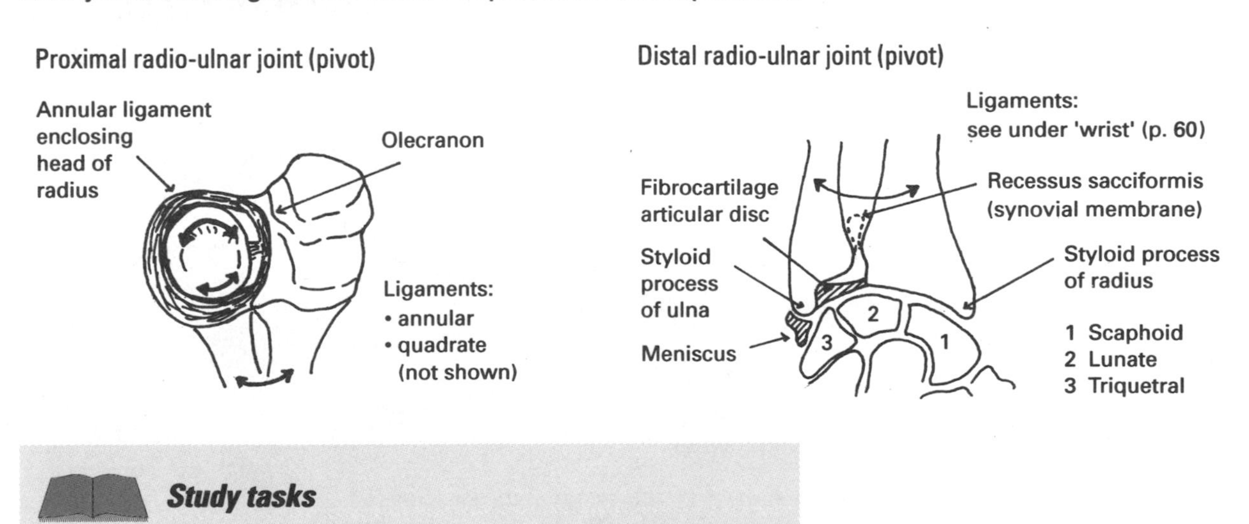

Study tasks

- Consider the details of the joints.
- Highlight the names of the features shown.
- Shade the bones in separate colours.

Supination rotates the forearm so that the palm of the hand faces anteriorly, while **pronation** rotates the forearm so that the palm faces posteriorly. The movements are best examined clinically with the elbow flexed and held into the side, in order to avoid shoulder rotation.

Supination

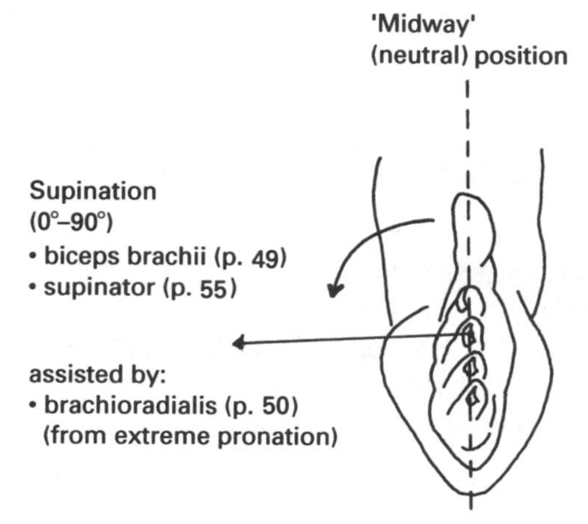

The supinating action of biceps brachii

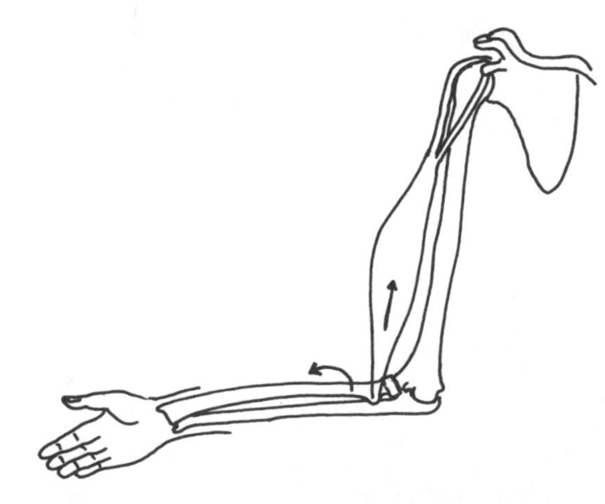

Study tasks

- Highlight the names of the muscles shown.
- Study the details of each muscle with reference to the page numbers given.
- In the case of biceps brachii consider why it is the strongest muscle of supination as well as an elbow and shoulder flexor.

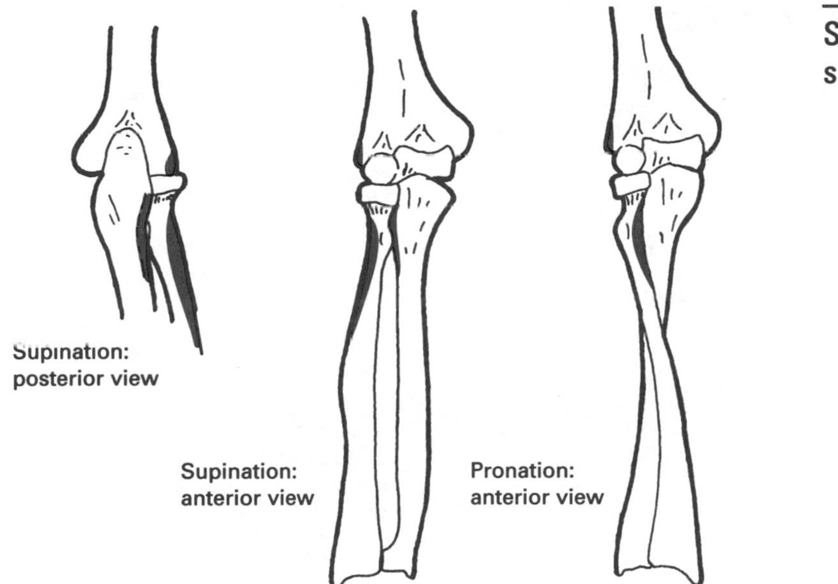

Supinator ('muscle of supination')

Attachments

- The lateral epicondyle of the humerus; the annular ligament and radial collateral ligament.
- The supinator fossa and crest of the ulna.
- The lateral aspect of the upper radius.

Nerve supply

Posterior interosseous nerve (C5, C6).

Function

- It supinates the forearm by 'unwinding' the radius from relative pronation: it is more effective when the elbow is extended.

Study tasks

- Shade lines between the attachment points, and consider the muscle function.
- Test the action of supinator on a colleague with arm and shoulder extended. Grasp their wrist and resist supination. Next test biceps by resisting supination with your colleague's elbow in flexion.
- Which muscle is the stronger? Deduce from anatomical observations rather than from palpation alone.
- Consider some practical applications of supination.

Pronation

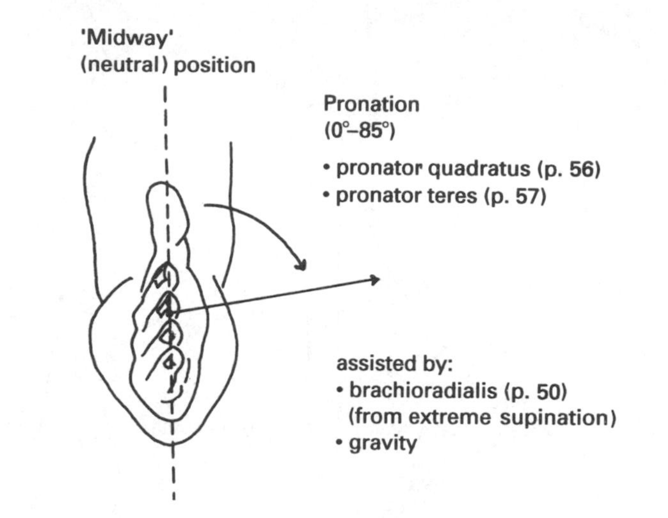

Study tasks

- Highlight the names of the muscles shown.
- Study the details of each muscle with reference to the page numbers given.

Pronator quadratus ('square-shaped muscle of pronation')

This is the most powerful muscle of pronation, and is a good example of the strong quadrilateral arrangement of parallel muscle fasciculi (p. 13). Notice how it also achieves mechanical advantage through its location distally near the wrist, and close to the distal radio-ulnar joint.

Attachments

- The distal anterior surface of the ulna.
- The distal antero-medial surface of the radius

Nerve supply

Median nerve (anterior interosseous branch) (C8, T1).

Function

- Pronation of the forearm

Study tasks

- Shade lines between the attachment points and consider the muscle function.
- Differentiate between the origin and insertion in this case, in order to achieve pronation.

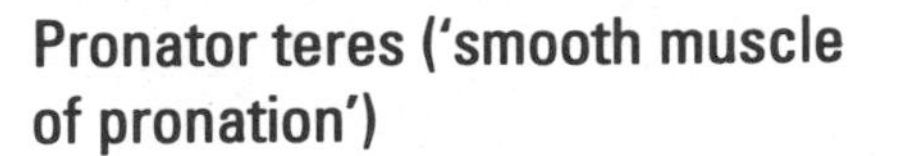

Pronator teres ('smooth muscle of pronation')

In supination

In pronation

Attachments

- The medial epicondyle and supracondylar ridge of the humerus.
- The coronoid process of the ulna.
- The middle of the lateral radius.

Nerve supply

Median nerve (C6, C7).

Functions

- Pronation of the forearm.
- Assists elbow flexion.

Study tasks

- Shade lines between the attachment points and consider the muscle functions.
- Why is this muscle unlikely to be as powerful as pronator quadratus?
- Identify the origin and insertion in order to produce pronation.
- Pronators teres and quadratus can be muscle tested separately, one with elbow fully flexed, and the other with a straighter arm. Which test would be preferable for each muscle and why?
- Is pronation stronger or weaker than supination and why? Are there any practical applications to support your view?

Chapter 6

The wrist

Introduction

The wrist is a functional region of small interlocking irregular bones, which accommodate the change in force and function from the gross movements of the arm towards the more precise and delicate movements of the hand.

The bones and joints

The individual carpal (wrist) bones are initially complex in name, shape and location, but may be split functionally into two rows: the distal carpal row and the proximal carpal row.

The bones and joints of the wrist (palmar surface)

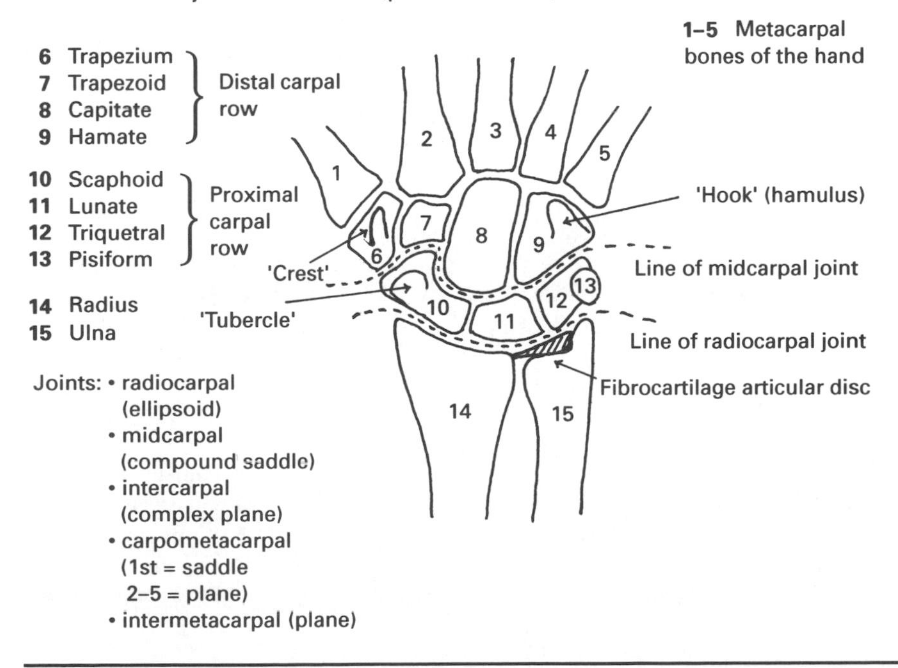

The wrist can be subdivided into two functional joints based on these rows.

- The radiocarpal joint between the head of the radius and the proximal row.
- The midcarpal joint between the proximal and distal rows. This is really a functional boundary, although Gray gives it the status of a 'compound sellar joint' (saddle joint) (*Gray's Anatomy* 1995).

Study tasks

- Shade the carpal bones in different colours and memorize the names.
- Highlight the names of the two functional joint lines described above.
- Obtain a bone specimen and identify the features shown.

The joints between the carpal bones (intercarpal joints) are all complex synovial plane joints, except for the radiocarpal joint which is synovial and ellipsoid. Like all synovial joints, they are strengthened by ligaments, and the number of bones and joint surfaces makes the overall situation complicated. However, the picture may be simplified to some extent by classifying the ligaments into three types:

- the ligaments between the bones (interosseous);
- the ligaments reinforcing the sides (collateral);
- the ligaments above and below (dorsal and palmar).

In addition to these, the **flexor retinaculum** is a strong fibrous structure (virtually an accessory ligament) which is attached to the crest of the trapezium and tubercle of the scaphoid on the lateral side, and to the 'hook' of the hamate and pisiform on the medial side. It converts the anterior carpal bones into a 'carpal tunnel' through which the flexor tendons of the fingers (and median nerve) pass.

Ligaments of the wrist

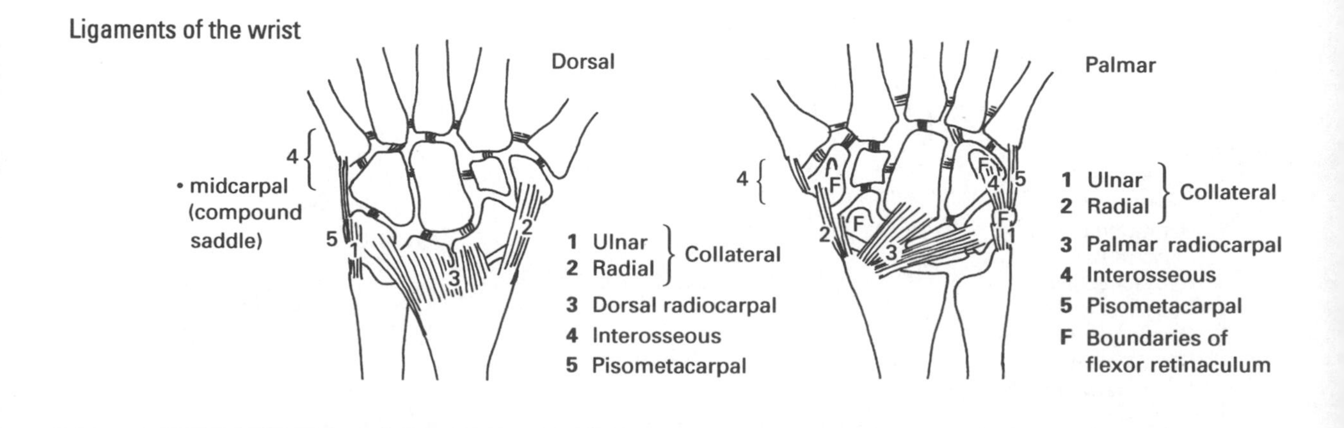

Study tasks

- Colour the ligaments, and highlight their names.
- Lightly mark the boundaries of the flexor retinaculum.

Movements

Apart from small adjustive gliding movements between the complex plane joints, the main wrist movements are described below.

Flexion

All the main flexor muscles of the wrist insert into the medial epicondyle of the humerus which is often referred to as the site of the 'common flexor tendon'. It is an area which is vulnerable to repetitive strain injuries if constant and sudden flexion movements occur. 'Golfer's elbow' (medial epicondylitis) is the term often used for repetitive strain injuries at this site even if the game has not been played!

The 'radialis' muscle will also produce radial deviation, or abduction, while the 'ulnaris' muscle will produce ulnar deviation or adduction.

Flexion of the wrist (mainly at midcarpal joint) (0°–85°)

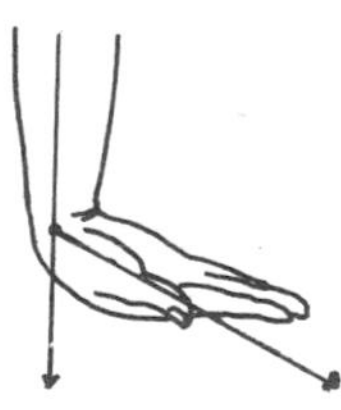

- flexor carpi radialis (p. 62)
- flexor carpi ulnaris (p. 63)
- palmaris longus (p. 64)

assisted by:
- flexor digitorum superficialis and profundus (pp. 84–85)
- flexor pollicis longus (p. 77)
- abductor pollicis longus (p. 80)

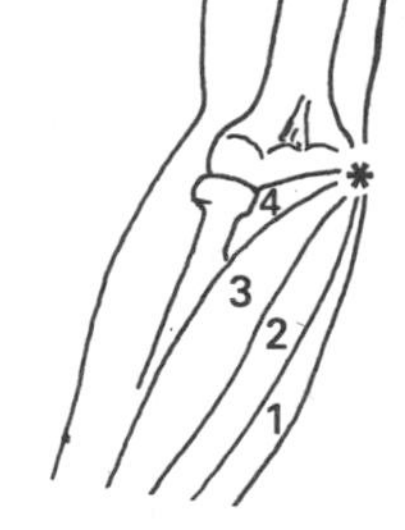

* Medial epicondyle – site of common flexor tendon

1 Flexor carpi ulnaris (p. 63)
2 Palmaris longus (p. 64)
3 Flexor carpi radialis (p. 62)
4 Pronator teres (p. 57)

Study tasks

- Highlight the names of the muscles shown.
- Shade the individual muscles which attach to the common flexor tendon, in separate colours.
- Study the details of each muscle with reference to the page numbers given.

Flexor carpi radialis ('flexor muscle of the wrist on the radial side')

Attachments

- ■ Common flexor tendon.
- ■ Base of second metacarpal.

Nerve supply

Median nerve (C6, C7).

Functions

- ♦ Flexion of the wrist.
- ♦ Abduction of the wrist.

Study tasks

- Shade lines between the attachment points and consider the muscle functions.
- Test the muscle by placing a colleague's wrist in mid-flexion with abduction, and resist full flexion. The muscle is superficial; the tendon can be observed lateral to palmaris longus (if present).

Flexor carpi ulnaris ('flexor muscle of the wrist on the ulnar side')

Attachments

- Common flexor tendon and medial olecranon and posterior ulna.
- Pisiform bone and its ligaments.

Nerve supply

Ulnar nerve (C7, C8).

Functions

- Flexion of wrist.
- Adduction of wrist.

Study tasks

- Shade lines between the attachment points and consider the muscle functions.
- Test the muscle by placing a colleague's wrist in mid-flexion with adduction, and resist full flexion. The muscle is superficial, but the tendon is less visible than flexor carpi radialis.

Palmaris Longus ('long muscle of the palm')

Note: This muscle is often absent on one or both sides.

Attachments

- Common flexor tendon.
- The palmar aponeurosis.

Nerve supply

Median nerve (C7, C8).

Function

- Flexion of the wrist and tightening of the palmar fascia.

Study tasks

- Shade a few lines between the attachment points.
- Test the muscle by placing a colleague's wrist in mid-flexion and resisting full flexion by exerting pressure across the middle of the palm. The tendon is superficial and is easily visible medial to flexor carpi radialis.

Extension

As with the flexor muscles, there is a common point of insertion on the lateral epicondyle of the humerus which is also vulnerable to repetitive strain, and the term 'tennis elbow' is often used to describe the lateral epicondylitis which occurs here. The 'radialis' and 'ulnaris' muscles also supply abduction and adduction, respectively.

Extension of the wrist (0°–85°) (mainly at the radiocarpal joint)

- extensor carpi radialis longus (p. 66)
- extensor carpi radialis brevis (p. 67)
- extensor carpi ulnaris (p. 68)

assisted by:
- extensor digitorum (p. 90)
- extensor digiti minimi (p. 91)
- extensor indicis (p. 91)
- extensor pollicis longus (p. 78)

* Lateral epicondyle – site of common extensor tendon

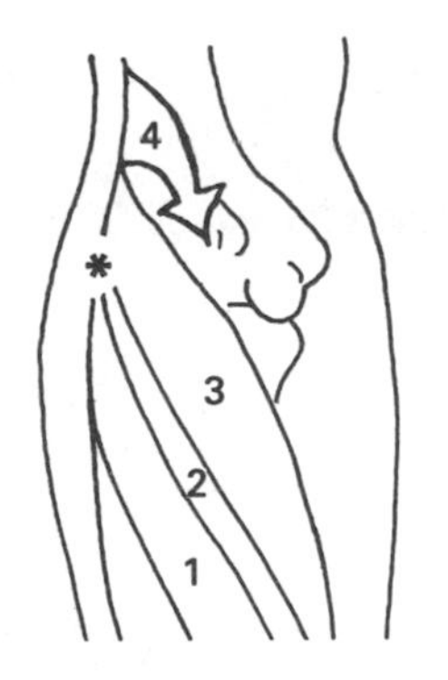

1 Extensor carpi ulnaris (p. 68)
2 Extensor digitorum (p. 90)
3 Extensor carpi radialis brevis (p. 67)
4 Site of extensor carpi radialis longus (p. 66) (schematic representation only)

Study tasks

- Highlight the names of the muscles shown.
- Shade the individual muscles which attach to the common extensor tendon, in separate colours.
- Study the details of each muscle with reference to the page numbers given.

Extensor carpi radialis longus ('long extensor muscle on the radial side of the wrist')

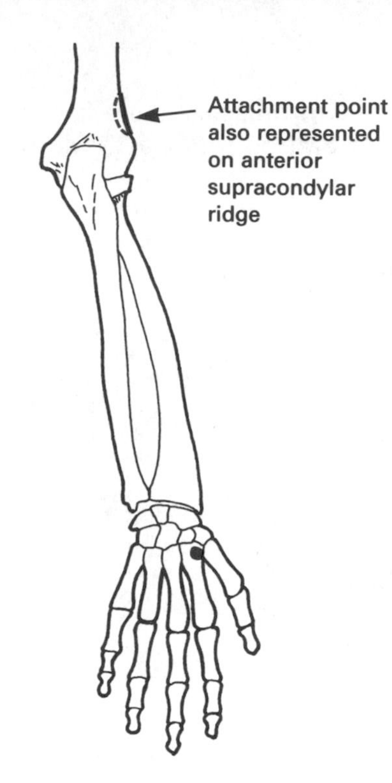

Attachments

- The lateral supracondylar ridge of the humerus (just above the common extensor tendon).
- The base of the second metacarpal.

Nerve supply

Radial nerve (C6, C7).

Functions

- Extension of the wrist.
- Abduction of the wrist.

Study tasks

- Shade lines between the attachment points and consider the muscle functions.
- Test this muscle with extensor carpi radialis brevis.

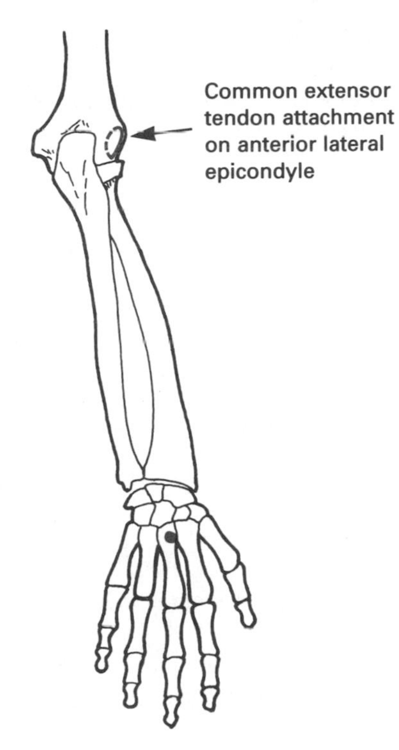

Extensor carpi radialis brevis ('short extensor muscle on the radial side of the wrist')

Attachments

- Common extensor tendon.
- Base of the third metacarpal.

Nerve supply

Posterior interosseous nerve (C7, C8).

Functions

- Extension of the wrist.
- Abduction of the wrist.

Study tasks

- Shade lines between the attachment points and consider the muscle functions.
- Test the radial extensor muscles by placing a colleague's wrist in slight extension with abduction. Turn the wrist into pronation and resist extension. The tendons of both extensor muscles can be palpated at the base of the second and third metacarpals. Palpate the attachment point at the common extensor origin.

Extensor carpi ulnaris ('extensor muscle of the wrist on the ulnar side')

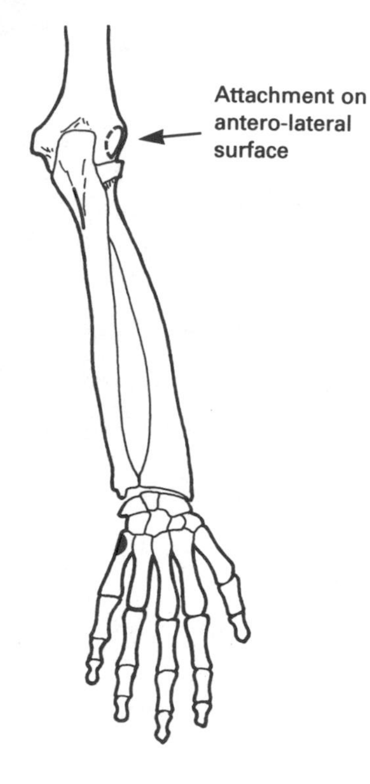

Attachments

- Common extensor tendon, and the border of posterior ulna.
- Base of the fifth metacarpal.

Nerve supply

Posterior interosseous nerve (C7, C8).

Functions

- Extension of the wrist.
- Adduction of the wrist.

Study tasks

- Shade lines between the attachment points and consider the muscle functions.
- Test the muscle on a colleague's wrist held in slight extension with strong adduction. Resist full extension.

Abduction

- flexor carpi radialis (p. 62)
- extensor carpi radialis longus and brevis (pp. 66–67)

assisted by:

- abductor pollicis longus (p. 80)
- extensor pollicis brevis (p. 79)

Abduction of the wrist (0°–15°) (mainly at the mid-carpal joint)

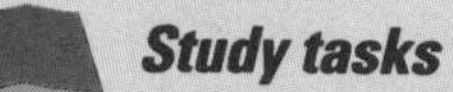

Study tasks

- Highlight the names of the muscles shown.
- Study the details of each muscle with reference to the page numbers given.

Adduction

Adduction of the wrist (0°–45°) (mainly at radiocarpal joint)

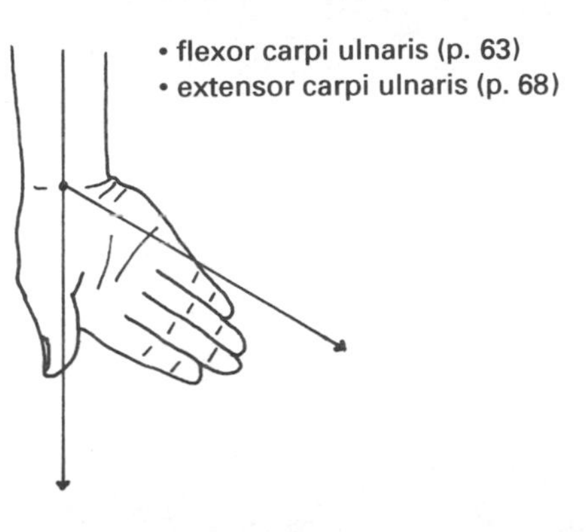

Study tasks

- Highlight the names of the muscles shown.
- Study the details of each muscle with reference to the page numbers given.
- Consider why the range of movement in the wrist is greater in adduction than in abduction.
- Consider the movement of circumduction with reference to the wrist and the muscles producing the movement.

Note: For supination/pronation movements see pp. 53–57.

Chapter 7

The hand

Introduction

The progress and achievements of human civilization are indebted to the versatility of the human hand.

Most movements of the upper limb are designed to move the hand into a position of best advantage; and once there, a wide variety of actions are possible. These can be described as 'grasping' and 'non-grasping' actions. The non-grasping actions include pushing, clapping, chopping, slapping, waving, praying, tickling and so on. But it is the grasping actions which display the real ingenuity of the hand, and these can be subdivided into four types of grip. (For more detailed analysis see Kapandji (1987).)

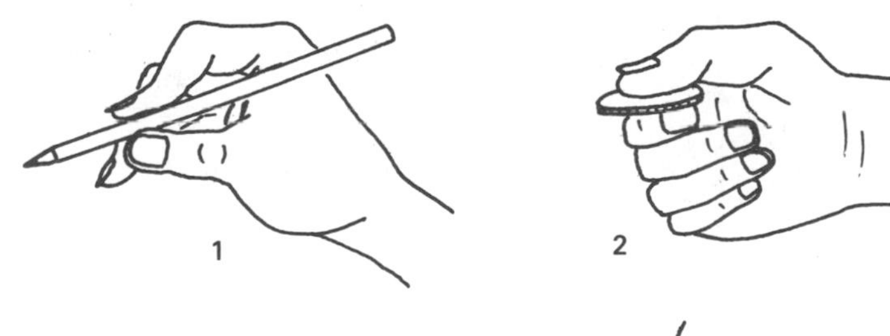

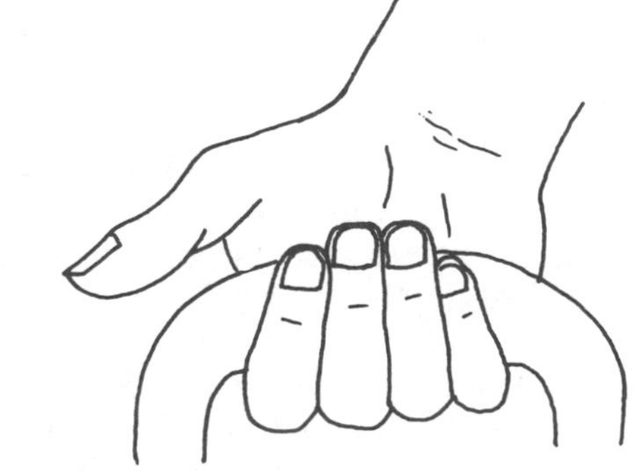

1. The chuck grip which uses at least two fingers and the thumb and operates like a drill chuck.
2. The pliers grip between thumb and index finger.
3. The ring grip which surrounds an object.
4. The hook grip which is less precise, and is the only one which does not rely heavily on the thumb.

The functional importance of the thumb cannot be overemphasized. The first carpometacarpal joint of the thumb is the most versatile joint in the hand and is medially rotated at rest. This means that the axes of movement in the thumb have been reorientated so that flexion/extension takes place in a plane parallel to the palm, and abduction/adduction occur in a plane perpendicular to this.

Movements of the hand

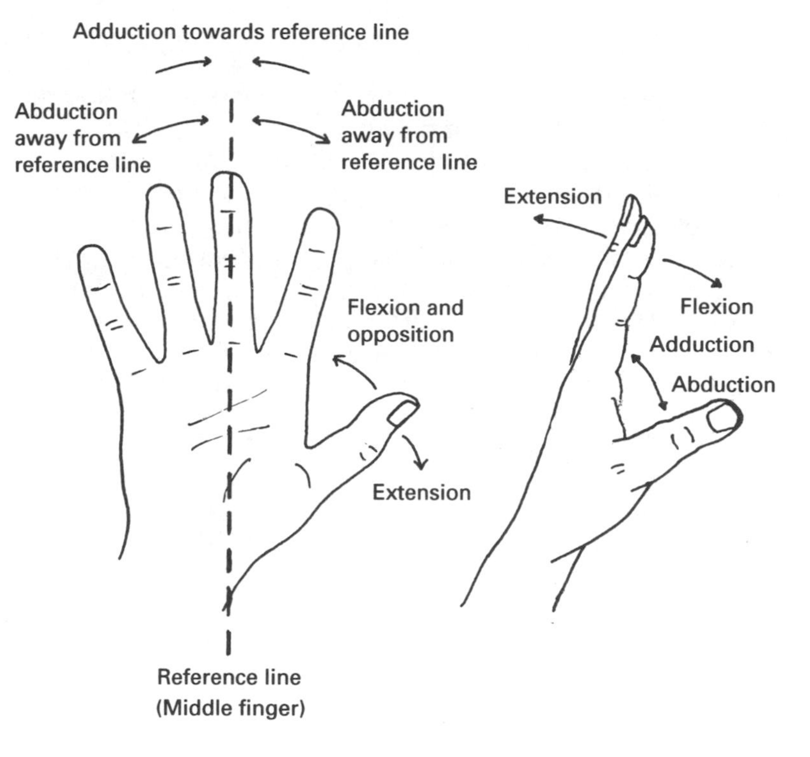

Study tasks

- Consider both the functional and the anatomical movements of both the fingers and the thumb. Ask a colleague to perform all movements actively. Pay particular attention to the movements of the thumb.
- How many actions of the hand is it possible to perform without using the thumb?

The joints of the fingers are essentially hinges capable of flexion/extension at the interphalangeal joints, with abduction/adduction at the metacarpophalangeal joints (knuckles) which are ellipsoid joints. Circumduction is also possible at the metacarpophalangeal joints.

There is, however, another feature of the hand which confers great versatility, and that is the fact that it can be 'arched'. This takes place along the line of the knuckles (metacarpophalangeal joints) and allows much of the enclosing action of the hand exemplified by the movement of opposition.

The arches of the foot receive considerable recognition but the arches of the hand are also functionally important: and of supreme importance in producing both the power and the precision of the human hand are the muscles.

The skin fits the hand rather like a rubber glove. The palmar surface has a thickened epidermal layer which allows better surface contact for grip, provides protective padding and has a rich supply of nerve endings for sensitivity. The palmar skin also has flexure creases (a fruitful source of speculation for fortune tellers) due to the anchorage of skin to deep fascia, and is a reflection of the repetitive movements of particular joints.

The dorsal surface is very different. It has a looser, thinner, hairy skin permitting full flexion of the fingers and opposition of the thumb.

The bones and joints

The bones of the hand form four functional arches which absorb the mechanical stress endured during manual work:

(i) a fairly rigid transverse carpal arch;
(ii) a mobile transverse metacarpal arch (at the knuckles);
(iii) a rigid longitudinal metacarpal arch formed by the shafts of the metacarpal bones;
(iv) a mobile interphalangeal arch formed by the fingers.

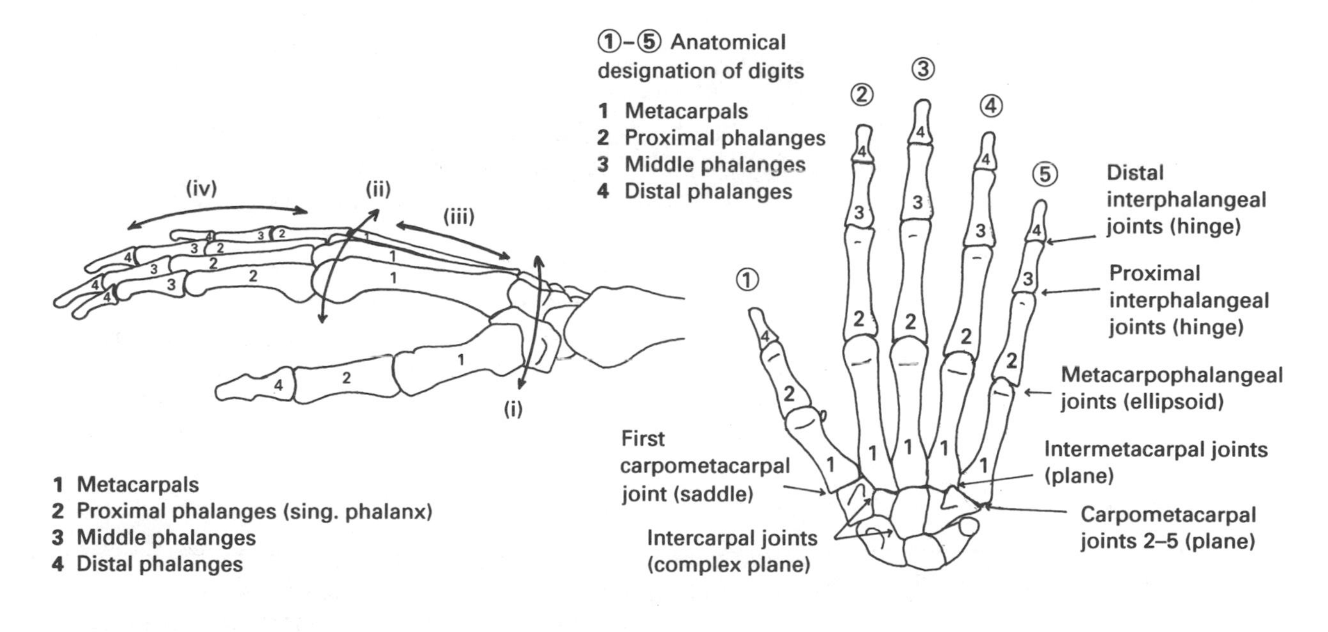

Study tasks

- Shade the carpal bones, the metacarpals and the phalanges, in different colours.
- Highlight the names of the joints.

The joints of the fingers are all synovial in kind, with some variation in type according to the movements required. In addition to the surrounding fibrous capsules, the ligaments which support the joints fall into three categories:

- the ligaments reinforcing the sides of the joint (collateral ligaments);
- the ligaments reinforcing the joint anteriorly and posteriorly (palmar and dorsal);
- the deep transverse metacarpal ligament.

The bones and ligaments of the fingers

M, metacarpal
P, phalanx

Sagittal view

1 Collateral ligaments
2 Palmar (volar) plates
3 Deep transverse metacarpal ligaments

Palmar view

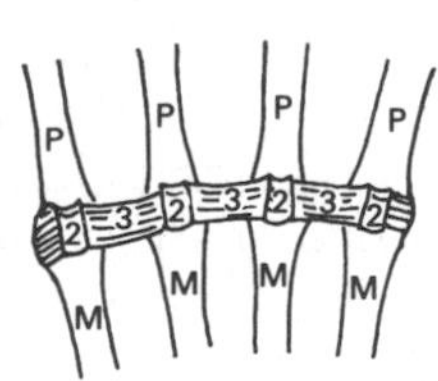

Study tasks

- Shade the ligaments in colour.
- Highlight the names of the ligaments.

Movements

The hand movements described earlier, which involve grasping and gripping with varying degrees of power and precision, are essentially compound movements involving coordination between the thumb and fingers. Many of the precision movements are controlled by muscles which are attached intrinsically within the hand. However, a number of the more powerful flexors and extensors take origin from the bones of the forearm.

Movements of the thumb

The two phalanges of the thumb are connected by the interphalangeal hinge joint which is only capable of flexion and extension. However, proximal to this is the first metacarpophalangeal joint which is ellipsoid allowing abduction/adduction movements, as well as flexion/extension. The thumb complex is then completed by the saddle-shaped first carpometacarpal joint which confers a degree of medial rotation in flexion and lateral rotation in extension, as well as allowing abduction/adduction to take place.

Flexion

Flexion of the thumb
(includes some medial rotation) (0°–50°)

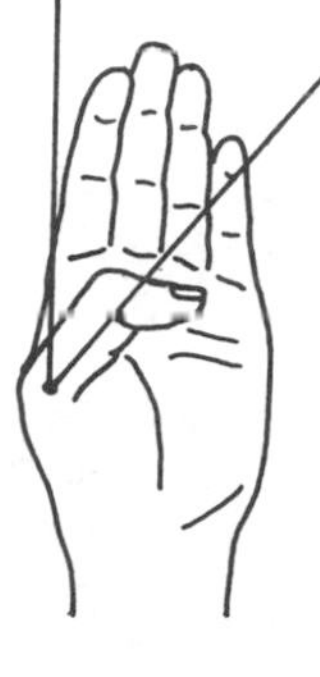

- flexor pollicis brevis (p. 75)
- opponens pollicis (p. 76)
- flexor pollicis longus (p. 77)

Study tasks

- Highlight the names of the muscles shown.
- Study the details of each muscle with reference to the page numbers given.

Flexor pollicis brevis ('short flexor muscle of the thumb')

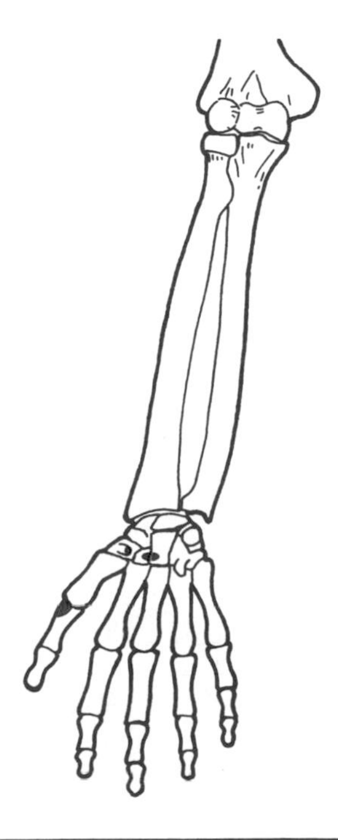

Attachments

■ Trapezium and flexor retinaculum; trapezoid and capitate.
■ Radial side of the proximal phalanx via a sesamoid bone.

Nerve supply

Median and ulnar nerves (C8, T1)

Function

♦ Flexion of the thumb.

Study tasks

- Shade lines between the attachment points and consider the muscle functions.
- Test the muscle by extending the entire thumb and opposing flexion midway along the proximal phalanx.

Opponens pollicis ('opposing muscle of the thumb')

Attachments

- ■ Trapezium and flexor retinaculum.
- ■ Radial aspect of first metacarpal.

Nerve supply

Median nerve (C8, T1).

Functions

- ♦ Flexion of the thumb.
- ♦ Opposition of the thumb.

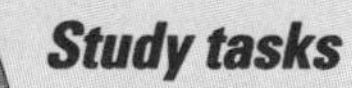

Study tasks

- Shade lines between the attachment points and consider the muscle functions.
- Test the muscle by opposing thumb and little finger. Try to push the first metacarpal into extension and palpate the degree of contraction in the thenar eminence.

Flexor pollicis longus ('long flexor muscle of the thumb')

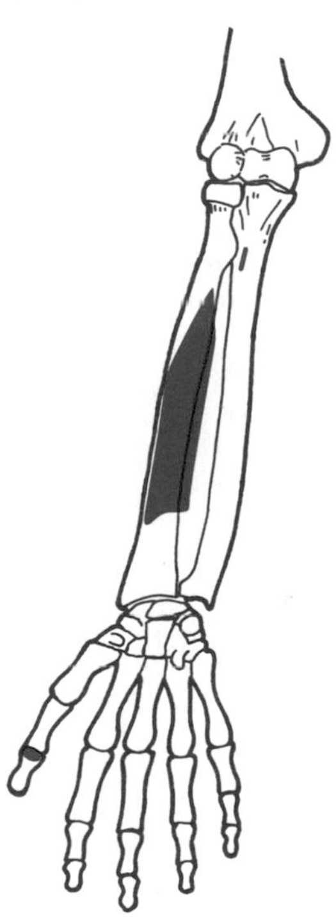

Attachments

- Middle half of anterior radius and interosseous membrane.
- Occasional head from anterior ulna.
- Base of the distal phalanx on palmer surface.

Nerve supply

Median nerve (anterior interosseous branch) (C8, T1).

Functions

- Flexion of all the thumb joints.
- Assists flexion of the wrist.

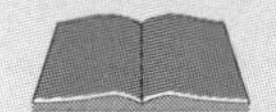

Study tasks

- Shade lines between the attachment points and consider why this muscle flexes all the thumb joints.
- Test the muscle by extending the entire thumb and resisting flexion at the distal phalanx.

Extension

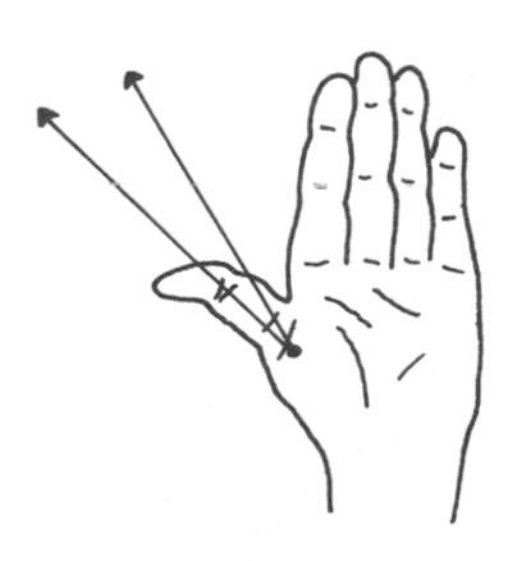

Extension of the thumb
(includes some lateral rotation (0°–50°) from the anatomical position)

- extensor pollicis longus (p. 78)
- extensor pollicis brevis (p. 79)
- abductor pollicis longus (p. 80)

Study tasks

- Highlight the names of the muscles.
- Study the details of each muscle with reference to the page numbers given.
- Shade the individual muscle tendons of the 'anatomical snuffbox' and highlight the muscle names.

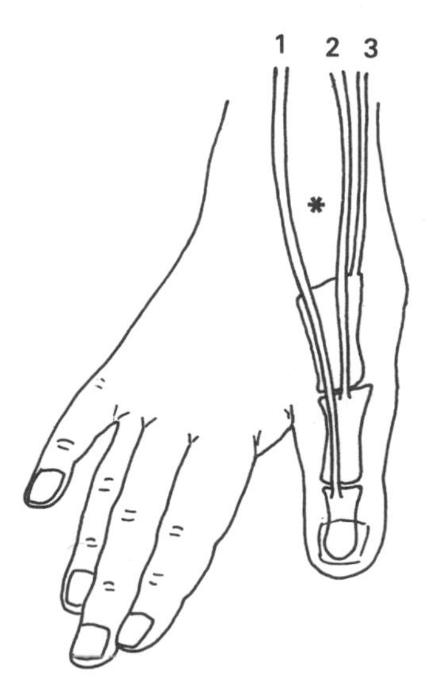

*The anatomical 'snuffbox'

1 Extensor pollicis longus (p. 78)
2 Extensor pollicis brevis (p. 79)
3 Abductor pollicis longus (p. 80)

Extensor pollicis longus ('long extensor muscle of the thumb')

Attachments

- Middle of the posterior ulna and interosseous membrane.
- Dorsal aspect of the base of the distal phalanx. The tendon passes via the dorsal tubercle on the radius (X) which acts as a pulley.

Nerve supply

Posterior interosseous nerve (C7, C8).

Functions

- Extension of all the thumb joints.
- Assists extension of the wrist.

Study tasks

- Shade lines between the attachment points and consider the muscle functions.
- Test the muscle by forcibly extending the entire thumb actively to its full extent. The tendon can be easily palpated as the medial boundary of the snuffbox. Also palpate the dorsal tubercle on the radius.

Extensor pollicis brevis ('short extensor muscle of the thumb')

Attachments

- Middle of posterior radius and interosseous membrane.
- Base of the proximal phalanx.

Nerve supply

The posterior interosseous nerve (C7, C8).

Functions

- Extension of the proximal phalanx of the thumb.
- May assist in extension of the wrist.

Study tasks

- Shade lines between the attachment points and consider the muscle functions.
- Test the muscle by flexing the distal phalanx and then extending the proximal phalanx of the thumb against resistance. Palpate the tendon as the more medial of the two in the lateral border of the snuffbox.
- Why do we flex the distal phalanx before we test this muscle?

Abductor pollicis longus ('long abductor muscle of the thumb')

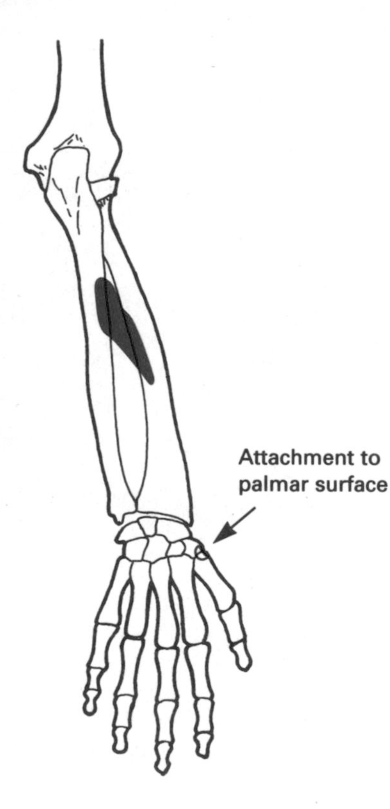

In spite of its name, this muscle is also an extensor of the thumb due to its attachment points.

Attachments

- Upper half of posterior aspect of the ulna, radius and interosseous membrane.
- Trapezium and base of the first metacarpal.

Nerve supply

Posterior interosseous nerve (C7, C8).

Functions

- Extension of the thumb.
- Abduction of the thumb.

Study tasks

- Shade lines between the attachment points and consider the muscle functions.
- Test the muscle with thumb flexed, and try to extend the metacarpal against resistance. The tendon is palpable as the most lateral in the anatomical snuffbox. The same tendon should be palpable if abduction is tested.

Abduction

Abduction of the thumb (0°–50°)

- abductor pollicis longus (p. 80)
- abductor pollicis brevis (p. 81)

Study tasks

- Highlight the names of the muscles shown.
- Study the details of each muscle with reference to the page numbers given.

Abductor pollicis brevis ('short abductor muscle of the thumb')

Attachments

- Scaphoid, trapezium and flexor retinaculum.
- Proximal phalanx, radial side.

Nerve supply

Median nerve (C8, T1).

Function

- Abduction of the thumb.

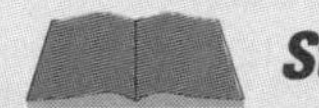

Study tasks

- Shade lines between the attachment points and consider the muscle function.
- Test the muscle by flexing the thumb until it is in line with the index finger, then resist abduction by pressing proximal phalanx towards the index finger. Observe contraction in the thenar eminence. This test is often used to examine the integrity of the median nerve.

Adduction

Adduction of the thumb

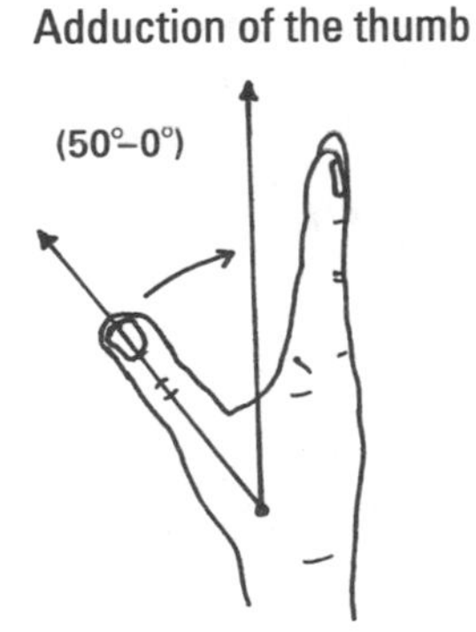

• adductor pollicis (p. 82)

Study tasks

- Highlight the name of the muscle shown.
- Study the details of the muscle overleaf.

Adductor pollicis ('adductor muscle of the thumb')

Attachments

- Anterior surface of third metacarpal (transverse head).
- Capitate, and bases of second and third metacarpals (oblique head).
- Base of proximal phalanx of the thumb via a sesamoid bone.

Nerve supply

Ulnar nerve (C8, T1).

Function

- Adduction of the thumb.

Study tasks

- Shade lines between the attachment points and consider the muscle function.
- Test the muscle by fully abducting the thumb and then resisting adduction. This test is often used to examine the integrity of the ulnar nerve.

Opposition

Opposition of the thumb

- opponens pollicis (p. 76)
- flexor pollicis brevis (p. 75)

Study tasks

- Highlight the names of the muscles shown.
- Study the details of each muscle with reference to the page numbers given.

Circumduction

Study tasks

- Highlight the individual movements which make up circumduction.
- Revise the muscles which produce each movement.
- Perform the individual movements on your own thumb in order to produce circumduction.

Circumduction of the thumb

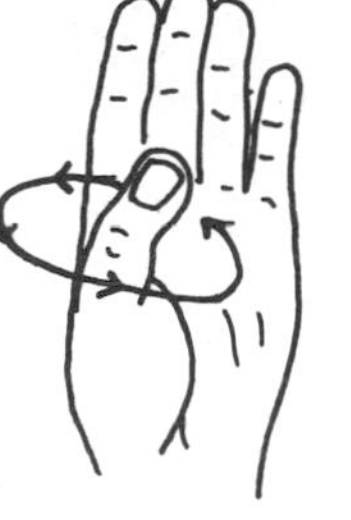

Movements of the hand

The configuration of the rigid and mobile arches allows the hand to cup, grip and generally mould around objects in either a powerful or a delicate way. This is achieved by the special mobility of the thumb, but also due to the greater laxity of the fourth/fifth intermetacarpal joints. This can be readily palpated by gripping the fifth metacarpal bone in your own hand and comparing its mobility with, for example, the second metacarpal. The cupping action is achieved by a combination of opposition between thumb and fourth and fifth fingers combined with adduction and flexion.

Flexion

Flexion takes place at the second–fifth metacarpophalangeal joints which are ellipsoid and possess a special supporting palmar ligament or plate; and at the interphalangeal joints of the fingers, which are hinge joints. The thumb may also flex.

Flexion of the hand

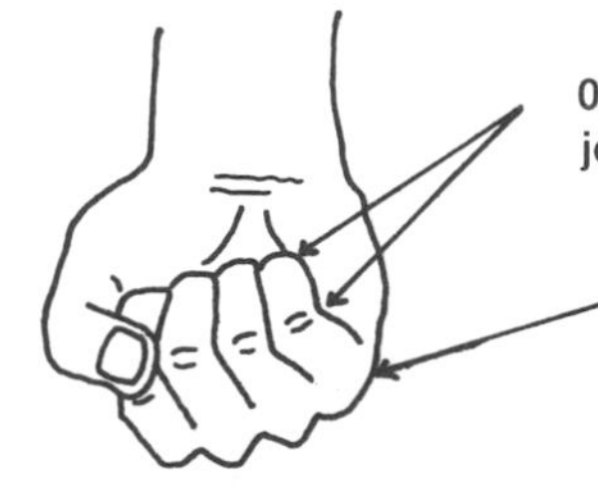

- flexor digitorum superficialis (p. 84)
- flexor digitorum profundus (p. 85)
- flexor pollicis longus (p. 77)
- flexor digiti minimi brevis (p. 86)

assisted by:
- lumbricals (pp. 86–87)
- interossei (pp. 87–89)
- see also 'flexion of the thumb' (p. 75)

Study tasks

- Highlight the names of the muscles shown.
- Study the details of each muscle with reference to the page numbers given.

Flexor digitorum superficialis (formerly 'sublimis') ('surface flexor muscle of the fingers')

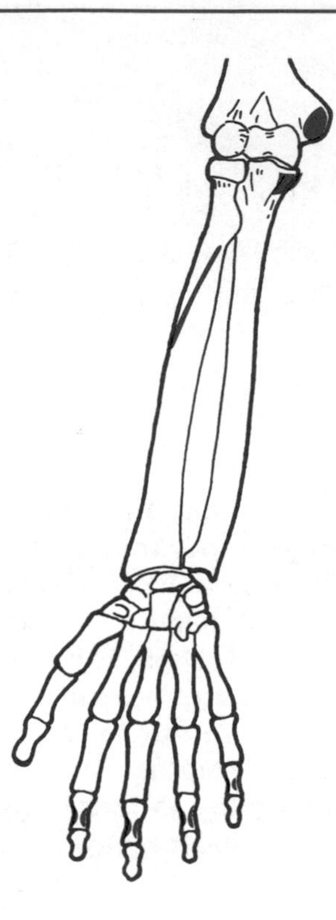

Attachments

- Common flexor tendon.
- Ulnar collateral ligament and coronoid tubercle.
- Middle part of anterior radius.
- Dual insertion on palmar surface of each middle phalanx 2–5.

Nerve supply

Median nerve (C7, C8, T1).

Functions

- Flexion of the proximal interphalangeal joints.
- Flexion of the metacarpophalangeal joints.
- Assists flexion of the wrist.

Study tasks

- Carefully shade lines between the attachment points, but first look at the diagram of the insertion arrangement under flexor digitorum profundus overleaf.
- Test the muscle by resisting flexion over the middle phalanges. Keep the distal phalanges straight.

Flexor digitorium profundus ('deep flexor muscle of the fingers')

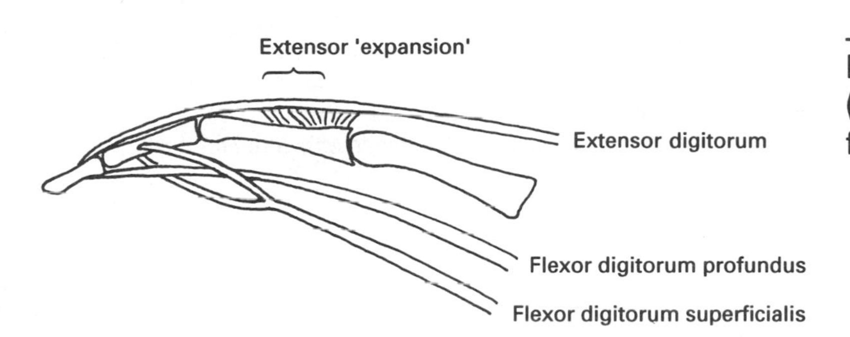

Sagittal view of tendon arrangement of digits 2–5 in the hand

Notice on the diagram above how the tendon of flexor digitorum profundus passes through the bifurcation (split) in flexor digitorum superficialis to insert deep into the distal phalanx: hence its name.

Attachments

- Upper part of the anterior ulna and interosseous membrane.
- Base of the second–fifth distal phalanges after passing through the bifurcation in the tendons of flexor digitorum superficialis.

Nerve supply

Median nerve (anterior interosseous branch) for lateral two tendons (C8, T1); ulnar nerve for medial two tendons (C7, C8).

Functions

- Flexion of all finger joints, but mainly the distal interphalangeal joints.
- Assists in wrist flexion.

Study tasks

- Colour the tendons of superficialis and profundus in separate colours on the diagram showing the tendon bifurcation.
- Shade lines between the muscle attachment points and consider the difference between the actions of superficialis and profundus.
- Self-test the muscle specifically by holding each finger firmly with the other hand and allowing only the distal interphalangeal joint to flex.

Flexor digiti minimi brevis ('short flexor of the little finger')

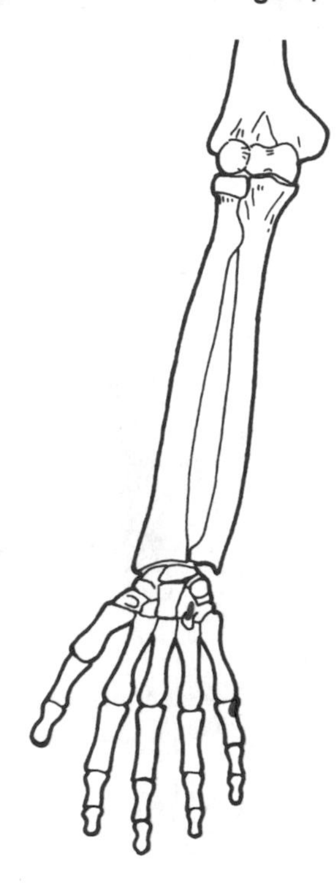

Attachments

- 'Hook' of the hamate and flexor retinaculum.
- Base of proximal phalanx of little finger with abductor digiti minimi (p. 92).

Nerve supply

Ulnar nerve (C8, T1).

Function

- Flexion of the little finger at the metacarpophalangeal joint.

Study tasks

- Shade lines between the attachment points and consider the muscle function.
- Self-test the muscle by placing a restraining finger from the opposite hand over the proximal phalanx of the little finger while flexion is attempted at the metacarpophalangeal joint.

The special action of the lumbricals and interossei

The lumbricals and interossei both flex the metacarpophalangeal joints, but their special *modus operandi* allows them to extend the interphalangeal joints at the same time, and thus produce the precise downstroke movements in writing and painting. The interossei also abduct and adduct the fingers.

The lumbricals ('worm-like muscles')

There are four lumbricals on each hand relating to the second to fifth fingers.

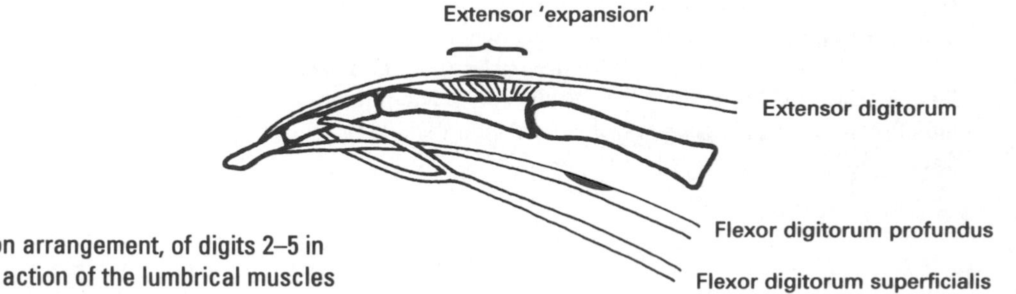

Sagittal view of tendon arrangement, of digits 2–5 in the hand to show the action of the lumbrical muscles

Attachments

- Origins on the tendons of flexor digitorum profundus.
- Insertions into the extensor aponeurosis of each finger on the lateral side. This merges with the tendon of extensor digitorum and is termed the extensor (dorsal) 'expansion' (*Gray's Anatomy*, 1980).

Nerve supply

Median nerve for the lateral two muscles (C8, T1); ulnar nerve for the medial two muscles (C8, T1).

Functions

- Flexion of the metacarpophalangeal joints with simultaneous extension of the interphalangeal joints.

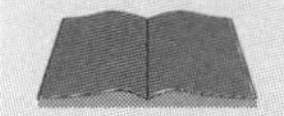

Study tasks

- Shade lines between the attachment points, and note how the insertion distal to the metacarpophalangeal joint will produce flexion, whilst the pulling effect on the extensor expansion draws the interphalangeal joints into extension.
- Refer back to the 'grasping' movements described (p. 71), and decide which ones can be refined into precision movements using the lumbrical muscles.
- Make some downstrokes with a pencil and observe movement in the joints described above.

The interossei ('the muscles between the metacarpal bones')

There are four pennate muscles, compact yet powerful, on both the palmar and dorsal sides of the hand. Their *modus operandi* is very similar to the lumbricals, but because of their side positioning they also abduct/adduct the fingers of the hand.

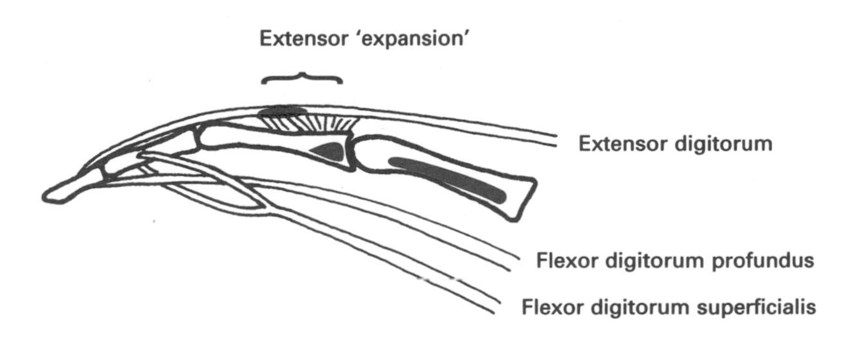

Sagittal view of tendon arrangement, of digits 2–5 in the hand to show the action of the interossei muscles

(i) The dorsal interossei

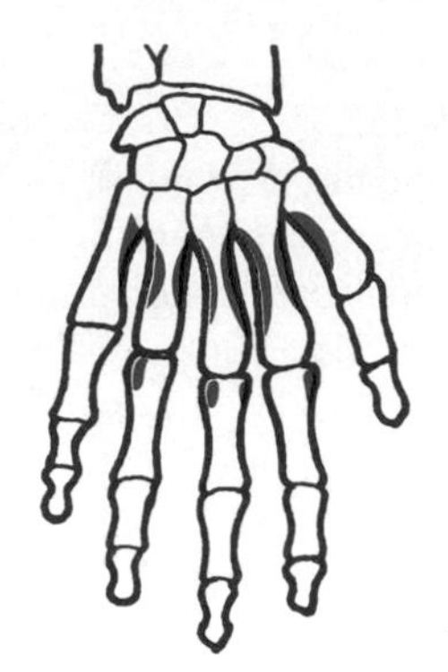

(Note:
Insertions into the
extensor expansion
not shown)

Attachments

- The adjacent sides of the dorsal shafts of the metacarpals.
- Insertions to the side of the base of the proximal phalanges as shown: then to the extensor expansion.

Nerve supply

Ulnar nerve (C8, T1).

Functions

- Flexion of the second–fourth metacarpophalangeal joints, with accompanying extension of the interphalangeal joints, as with the lumbricals.
- Abduction of the fingers away from the middle finger (p. 92). (The thumb and little finger have their own abductor muscles.)

Study tasks

- Shade lines between the attachment points in the first diagram, to illustrate the same action as the lumbricals.
- Shade lines between the attachment points in the second diagram to show the action of abduction.

Attachments

- Medial aspect of first metacarpal, and the anterior aspect of second, fourth and fifth metacarpals.
- Insertions to the side of the base of the relevant proximal phalanx as shown, and then to the extensor expansion.

Nerve supply

Ulnar nerve (C8, T1).

Functions

- Flexion of the second–fourth metacarpophalangeal joints with accompanying extension of the interphalangeal joints, as with the lumbricals.
- Adduction of the fingers and thumb towards the third finger.

Study tasks

- Shade lines between the attachment points and consider the muscle functions.
- Check that you understand the flexion/extension actions with reference to the sagittal diagram of the interossei.

(ii) The palmar interossei

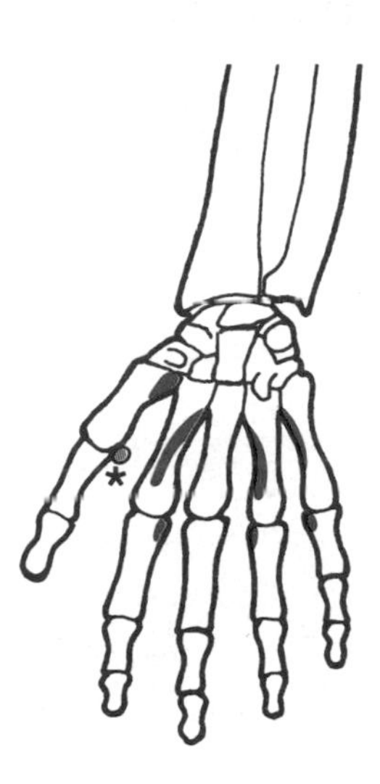

(Note:
Insertions into the extensor expansion not shown.)

* Insertion of first palmar interosseus muscle into a sesamoid bone.

Extension

Extension of the hand (0°–40° involving wrist extension and finger abduction)

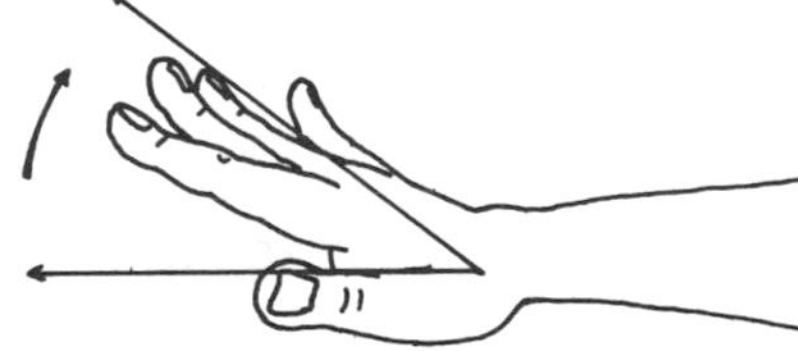

- extensor digitorum (p. 90)
- extensor indicis (p. 91)
- extensor digiti minimi (p. 91)
- extensor pollicis longus and brevis (p. 78, 79)

Study tasks

- Highlight the names of the muscles shown.
- Study the details of each muscle with reference to the page numbers given.

Extensor digitorum ('extensor muscle of the fingers')

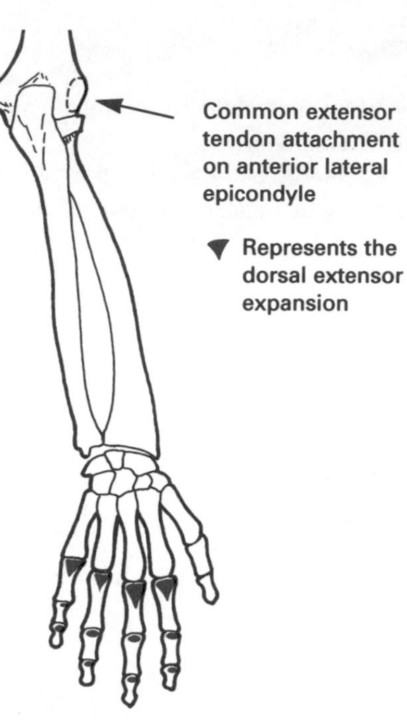

Attachments

- The common extensor tendon.
- The extensor expansion.
- Tendinous 'slips' to the proximal, middle and distal phalanges of each finger.

Nerve supply

Posterior interosseous nerve (C7, C8).

Functions

- Extension of the second to fifth fingers.
- Assists extension of the wrist.

Study tasks

- Shade lines between the attachment points and consider the muscle functions.
- Self-test the muscle by fully extending all fingers. Note how abduction also takes place. Why should this be so?
- Palpate the aponeurosis of the extensor expansion by pressing the middle finger (the finger between the index and 'ring' finger) against the tip of the thumb. The extensor expansion can be palpated as a ridge running along the medial edge of the proximal phalanx.

Extensor indicis ('extensor muscle of the index finger')

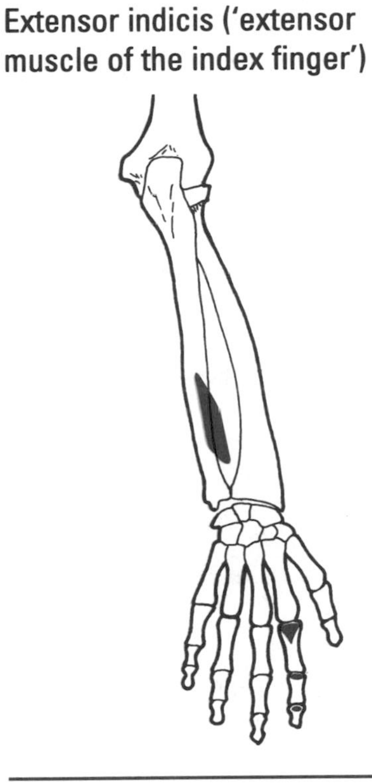

Attachments

- Posterior lower ulna and interosseous membrane.
- Extensor attachments of the index finger as for extensor digitorum (p. 90).

Nerve supply

Posterior interosseous nerve (C7, C8).

Function

- Extension of the index finger.

Study tasks

- Shade lines between the attachment points and consider the muscle function.
- Self-test the muscle by extending all fingers fully and adding resistance to the index finger.

Extensor digiti minimi ('extensor muscle of the little finger')

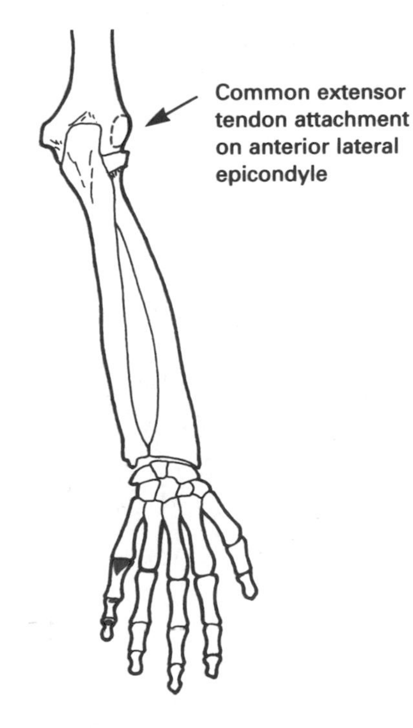

Attachments

- The common extensor tendon.
- Extensor attachments of the little finger as for extensor digitorum (p. 90).

Nerve supply

Posterior interosseous nerve (C7, C8).

Function

- Extension of the little finger.

Study tasks

- Shade lines between the attachment points and consider the muscle function.
- Self-test the muscle by extending the little finger against resistance.

Abduction

Abduction of the hand (accompanies extension)

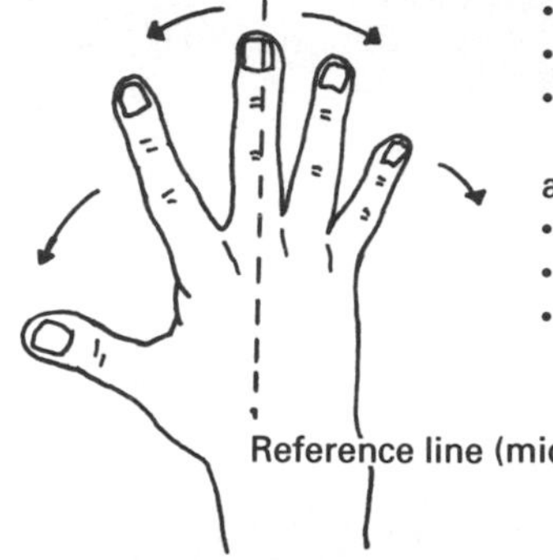

- dorsal interossei (p. 88)
- abductor digiti minimi (p. 92)
- abductor pollicis brevis (p. 81)

assisted by:

- extensor digitorum (p. 90)
- extensor indicis (p. 91)
- extensor digiti minimi (p. 91)

Study tasks

- Highlight the names of the muscles shown.
- Study the details of each muscle with reference to the page numbers given.
- Why is abduction of the fingers impossible if the metacarpophalangeal joints are flexed?

Abductor digiti minimi ('abductor muscle of the smallest finger')

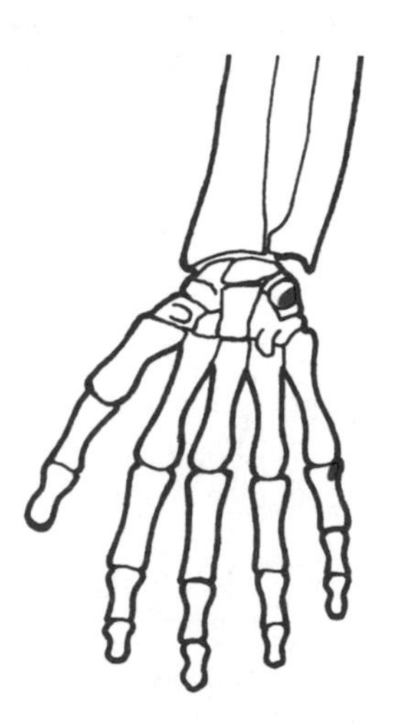

Attachments

- Pisiform carpal bone and flexor retinaculum.
- Base of the proximal phalanx of the little finger, on the ulnar side.

Nerve supply

Ulnar nerve (C8, T1).

Function

- Abduction of the little finger.

Study tasks

- Shade lines between the attachment points and consider the muscle function.
- Self-test the muscle by abducting the little finger in extension, against resistance.

Adduction

Adduction of the hand (accompanies flexion)

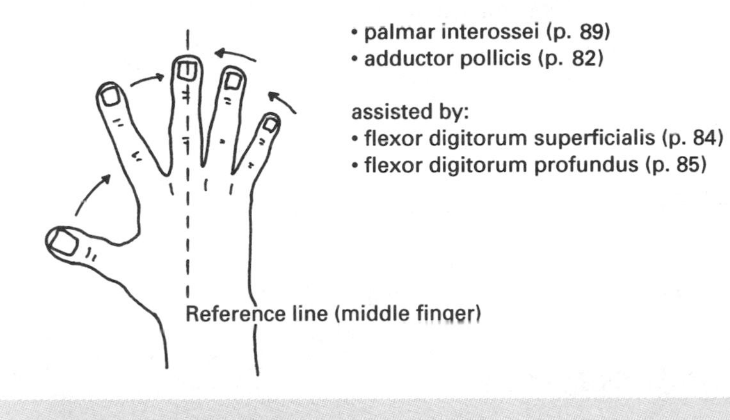

Study tasks

- Highlight the names of the muscles shown.
- Study the details of each muscle with reference to the page numbers given.

Opposition

Opposition of the little finger

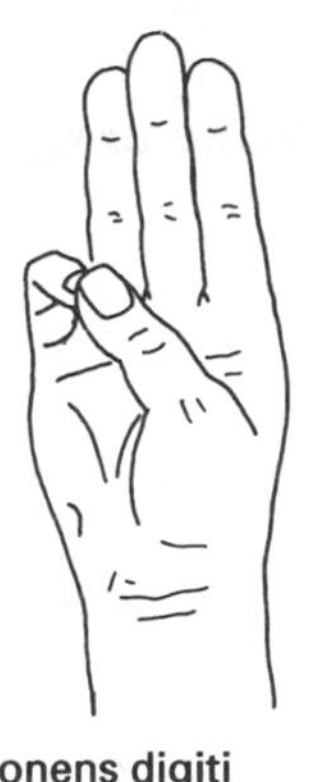

• opponens digiti minimi (p. 94)

Opposition of the fingers towards the opposing thumb can be adequately performed by using the normal finger flexor muscles and relying on the thumb to provide the dynamic opposition movement (p. 82). However, the efficiency of the movement is enhanced by the presence of a small muscle of opposition in the hypothenar eminence, attached to the little finger.

Opponens digiti minimi ('opposing muscle of the little finger')

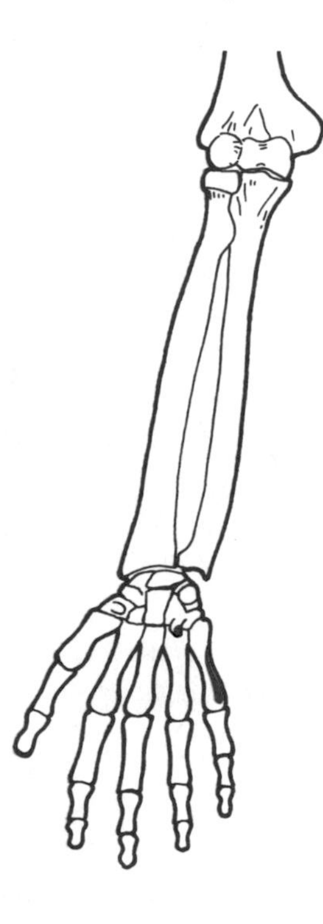

Attachments

- 'Hook' of the hamate and flexor retinaculum.
- Ulnar margin of the fifth metacarpal.

Nerve supply

Ulnar nerve (C8, T1).

Functions

- Opposition of the little finger towards the opposing thumb.
- Helps to 'cup' the palm of the hand.

Study tasks

- Shade lines between the attachment points and consider the muscle functions.
- Self-test the muscle by firmly opposing thumb and little finger. Notice also the 'cupping' effect in the palm of the hand.

Note: The 'cupping' effect is also enhanced by the action of a superficial muscle, **palmaris brevis**. This is a thin quadrilateral muscle passing from the flexor retinaculum and palmar aponeurosis to the dermis of the skin, on the ulnar side of the hand. It puckers the skin on the ulnar side, and thereby accentuates the hollow in the palm. The nerve supply of this muscle is the superficial branch of the ulnar nerve (C8, T1).

Chapter 8

The hip and pelvis

Introduction

The upper and lower limbs have basic similarities in design. They are both attached to the trunk by a specialized bony girdle; the free limbs consist of a single proximal bone (humerus and femur); this is continued distally by paired intermediate bones (radius/ulna and tibia/fibula); the limbs are 'completed' by a hand or foot.

However, the functional requirements are very different. The principal function of the non-weight-bearing upper limb is to orientate the hand so that its specialized movements may be performed. In the case of the lower limb, the principal functions are to provide weight-bearing support and locomotion for the upper body. In comparison with the upper limb, bones in the lower limb are larger; pronation/supination in the leg is not required (cf. radius/ulna); and general movements are grosser.

The hip and glenohumeral joints are both synovial ball-and-socket joints, but the hip is weight-bearing, and therefore has a deeper socket. This has both advantages and disadvantages. The hip is at once more stable, but is less mobile, and more susceptible to degenerative change.

The bones and joints

Bones and joints of the hip and pelvis

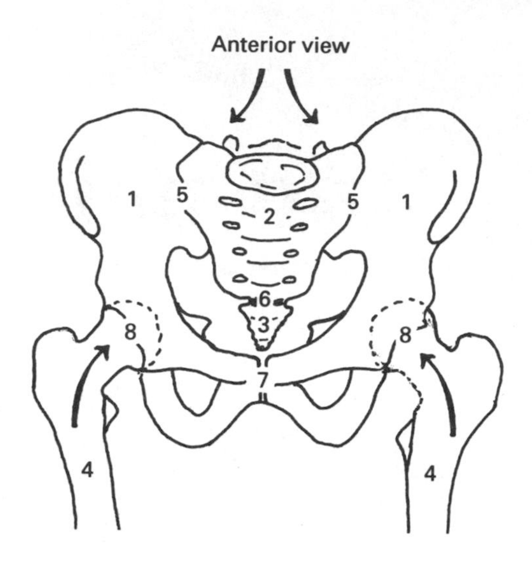

Posterior view

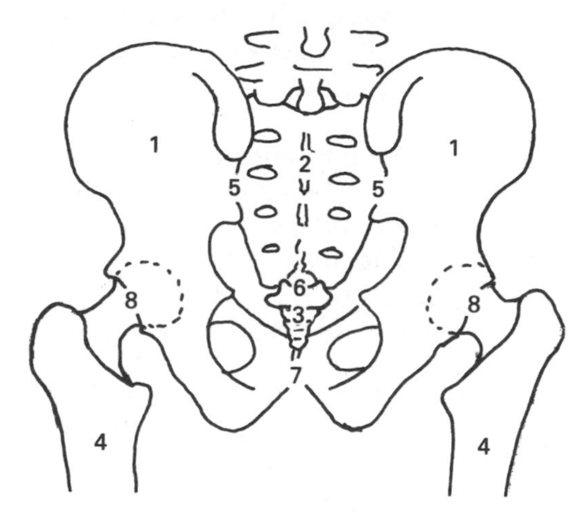

Bones

1. Innominate
2. Sacrum
3. Coccyx
4. Femur

Joints

5. Sacro-iliac joint (synovial)
6. Sacro-coccygeal joint (2° cartilaginous)
7. Pubic symphysis (2° cartilaginous)
8. Coxal (hip) (ball and socket)

↓↓ Weight-bearing forces ↑↑

Study tasks

- Shade the femur, sacrum, coccyx and innominate bones in separate colours.
- Highlight the names of the joints as indicated.

The hip joint consists of the spheroidal head of the femur which articulates with the acetabulum (the socket) of the pelvis. The pelvis forms a bony ring which serves a number of purposes, including the protection of abdominal viscera and obstetric functions in the female. However, its shape also helps to receive and transmit forces from above and below and the hip joint is integral to this.

The bony pelvic ring comprises two halves, and each 'half' or 'innominate' bone consists of three fused bones: the ilium (iliac bone), the ischium (ischial bone) and the pubis (pubic bone). These three bones meet at the acetabulum, are united by cartilage in the young, but ossify in the adult.

The right innominate bone (lateral aspect)

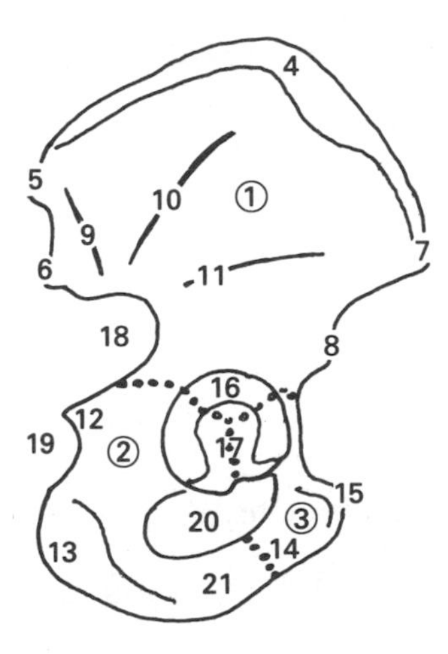

① ilium
② ischium
③ pubis

Limits of iliac, pubic and ischial bones

4 Iliac crest
5 Posterior superior iliac spine
6 Posterior inferior iliac spine
7 Anterior superior iliac spine
8 Anterior inferior iliac spine
9 Posterior gluteal line
10 Anterior gluteal line
11 Inferior gluteal line
12 Ischial spine
13 Ischial tuberosity
14 Pubic ramus
15 Pubic tubercle
16 Acetabulum
17 Acetabular notch
18 Greater sciatic notch
19 Lesser sciatic notch
20 Obturator foramen
21 Ischial ramus

Study tasks

- Shade each of the three parts in separate colours.
- Highlight the names of the bone features indicated.

The two innominate halves are united by the wedge-shaped sacrum posteriorly, which forms the two sacro-iliac joints, but directly by the pubic symphysis joint anteriorly.

Note: The sacro-iliac joints do not move voluntarily under active muscular control. They permit small adjustive movements during physical activity, and manipulative therapy often seeks to move the joints passively for treatment purposes. (See, for example, Hartman (1990).)

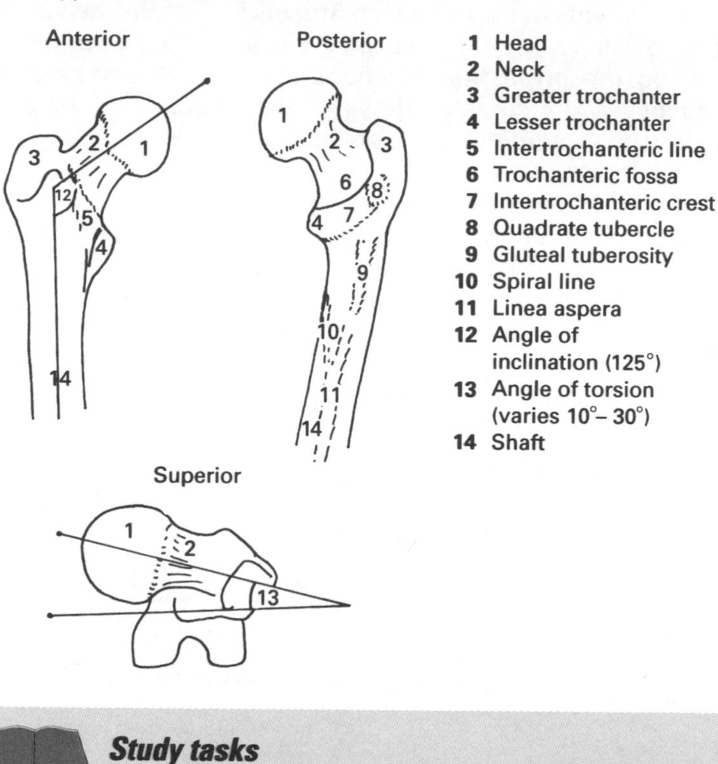

Study tasks

- Highlight the names of the bone features shown, including the angles of inclination and torsion.
- Obtain bone specimens and identify all the features indicated.

The weight-bearing functions of the hip joint are strikingly represented in the bone structure of the femur (thigh bone) and acetabular socket of the innominate (hip bone). The necessary width of the pelvis means that the femur possesses a much longer bony neck than the humerus. The shaft is then pitched at an angle (the angle of inclination) which allows the knees and feet to return closer to the midline. Also present is an angle between the neck of the femur and the frontal plane (the angle of torsion).

The power of the muscles which act on the hip is attested to by the size of the bony tubercles or trochanters to which the muscles attach. There are also well-marked lines of attachment, such as the linea aspera and intertrochanteric lines, on or near the shaft. In addition, the lines of force which pass through the hip are represented by well-defined linear trabeculae, visible in the inner spongy bone.

Ligaments of the hip and pelvis (anterior)

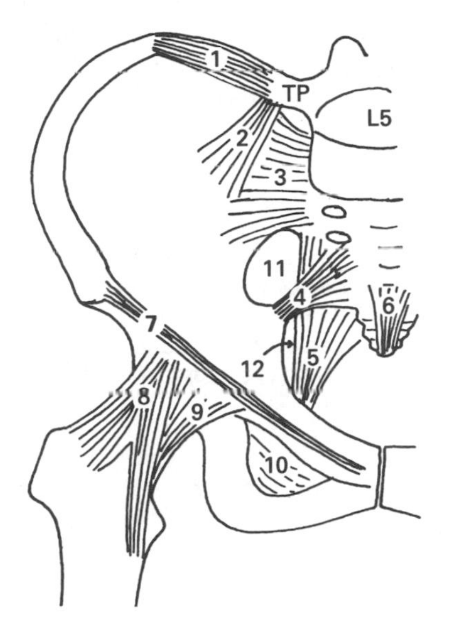

1 Iliolumbar
2 Lumbosacral
3 Ventral sacro-iliac
4 Sacrospinous
5 Sacrotuberous
6 Ventral sacrococcygeal
7 Inguinal
8 Iliofemoral
9 Pubofemoral
10 Obturator membrane
11 Greater sciatic foramen
12 Lesser sciatic foramen

TP Transverse process of L5

Ligaments of the hip and pelvis (posterior)

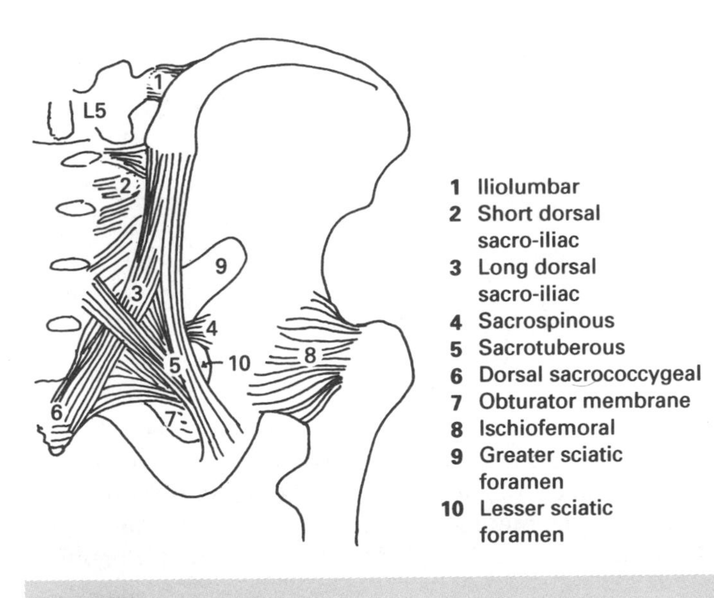

Study tasks

- Study the details of each joint, and shade the ligaments in colour.
- Highlight the names of the features shown.

The muscles

As in the case of the shoulder girdle (Ch. 4), a number of the superficial muscles can, with practice, be identified by observation and palpation. Before proceeding, it is worth noting the surface distribution of the muscles covering the hip, pelvis and thigh.

The superficial muscles (anterior)

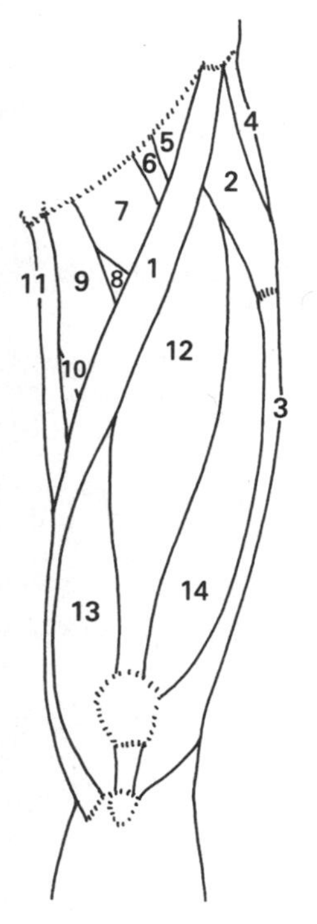

1 Sartorius (p. 105)
2 Tensor fasciae latae (p. 114)
3 Iliotibial tract
4 Gluteus medius (p. 112)
5 Iliacus (p. 103)
6 Psoas major (p. 102)
7 Pectineus (p. 103)
8 Adductor brevis (p. 116)
9 Adductor longus (p. 115)
10 Adductor magnus (p. 116)
11 Gracilis (pp. 105–106)
12 Rectus femoris (p. 104)
13 Vastus medialis (p. 135)
14 Vastus lateralis (p. 137)

The superficial muscles (posterior)

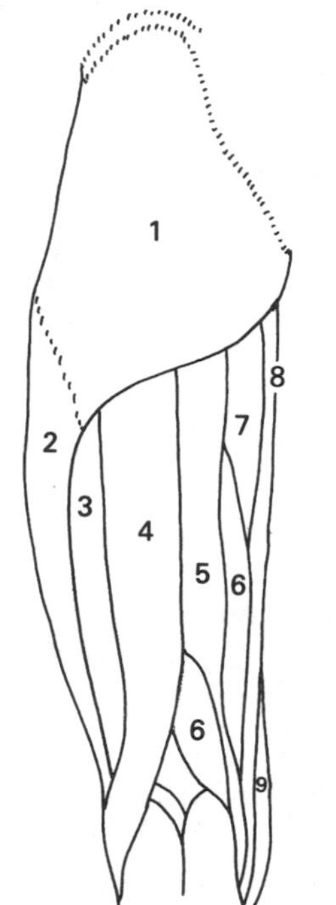

1 Gluteus maximus (p. 108)
2 Iliotibial tract
3 Vastus lateralis (p. 137)
4 Biceps femoris (p. 109)
5 Semitendinosus (p. 110)
6 Semimembranosus (p. 111)
7 Adductor magnus (p. 116)
8 Gracilis (pp. 105–106)
9 Sartorius (p. 105)

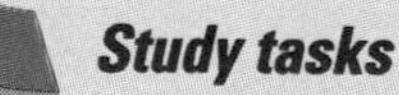

Study tasks

- Shade the muscles shown in separate colours.
- Highlight the names of the muscles.

Movements of the hip

The ball-and-socket joint of the hip allows flexion and extension through a transverse axis; abduction and adduction through an antero-posterior axis; medial and lateral rotation through a longitudinal axis; and the combined movement of circumduction.

Flexion

Flexion brings the anterior surface of the thigh closer to the trunk. The range of movement is increased if the knees are bent, and this is increased even further if the knees are drawn up to the chest.

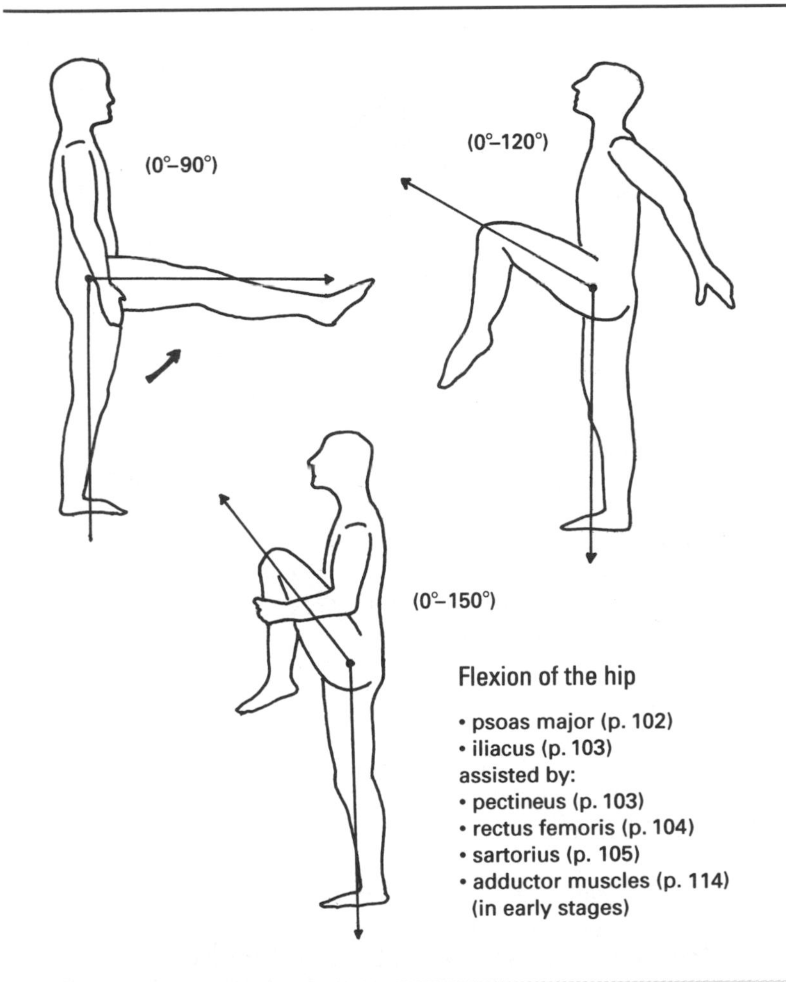

Flexion of the hip

- psoas major (p. 102)
- iliacus (p. 103)

assisted by:

- pectineus (p. 103)
- rectus femoris (p. 104)
- sartorius (p. 105)
- adductor muscles (p. 114) (in early stages)

Study tasks

- Demonstrate the varying degrees of hip flexion either on yourself or a colleague.
- Explain why these variations occur.
- Highlight the names of the muscles producing flexion.
- Study the details of each muscle with reference to the page number given.

Psoas major ('larger loin muscle')

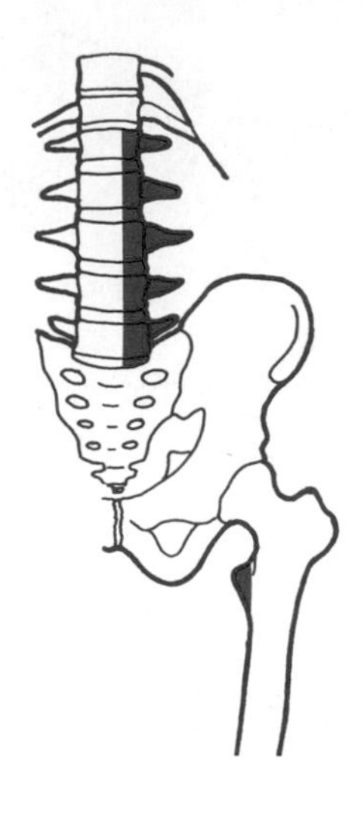

Attachments

- TPs of L1–5; vertebral bodies and discs of L1–5.
- Lesser trochanter of the femur.

Nerve supply

Ventral rami of L1, L2, L3.

Functions

- Flexion of the hip.
- Some lateral rotation of the femur.
- Unilaterally it assists sidebending of the lumbar spine, with some flexion.
- Bilaterally, flexion of the lumbar spine.
- Bilaterally, in the standing position its anterior attachments maintain, and even exaggerate the lumbar lordosis.

Study tasks

- Shade lines between the attachment points and consider the muscle functions.
- Why is a rigid application of the concept of origin and insertion not appropriate to this muscle?
- Observe the attachment points on a skeletal specimen to appreciate the function of lateral rotation, and the role of maintenance of the lumbar lordosis.
- Test the muscle by placing a colleague in the supine position with their thigh and leg held at 45° and in slight lateral rotation. Support their leg at the knee, but add slight resistance to the medial aspect of their thigh.

Iliacus ('muscle of the ilium')

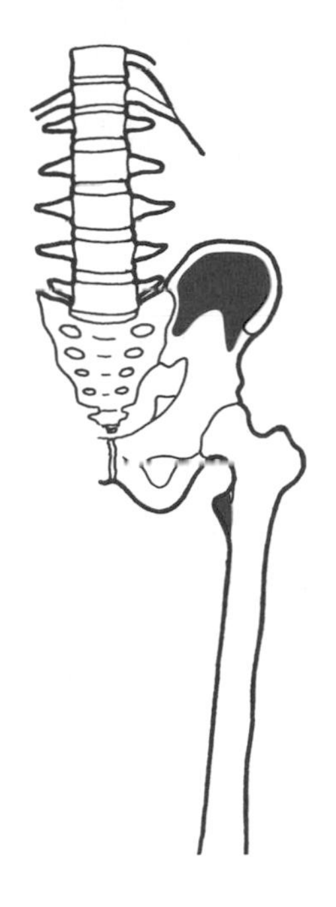

This muscle is so integral to hip flexion with psoas major that they are often grouped together as 'ilio-psoas'.

Attachments

- Upper part of the iliac fossa and inner lip of the iliac crest.
- Lesser trochanter of the femur (with psoas major).

Nerve supply

Femoral nerve (L2, L3).

Functions

- Flexion of the hip.
- Anterior movement of the pelvis.

Study tasks

- Shade lines between the attachment points and consider the muscle functions.
- How does the difference in attachment points explain its more limited role compared with psoas major?
- Test the muscle as for psoas major.

Pectineus ('comb or lyre-shaped muscle')

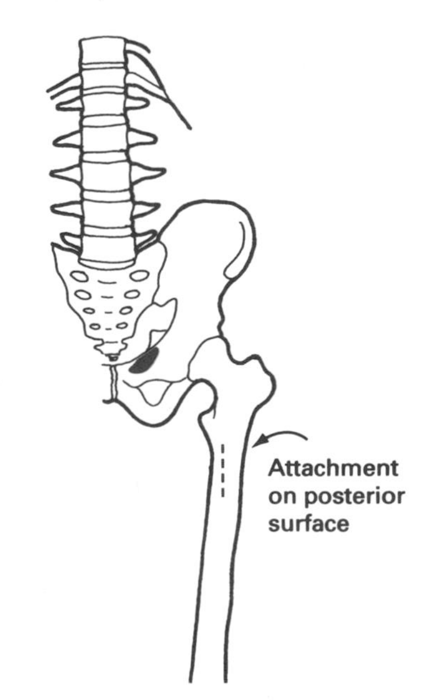

Attachments

- Above the pubic tubercle.
- Between the lesser trochanter and linea aspera.

Nerve supply

Femoral nerve (L2, L3) and sometimes a branch from the obturator nerve.

Functions

- Assists flexion of the hip.
- Adduction of the hip.

Study tasks

- Shade lines between the attachment points and consider the muscle functions.
- See also under Adduction of the hip (p. 114).

Rectus femoris ('thigh-raising muscle')

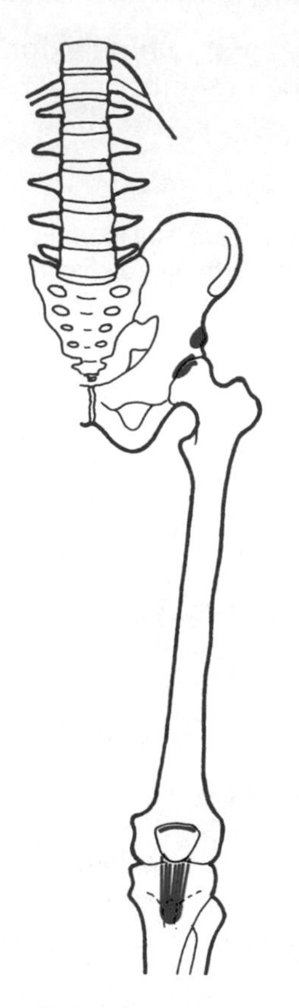

Rectus femoris is part of the quadriceps group of muscles which collectively attach to the ligamentum patellae and tibial tuberosity, and therefore produce knee extension (p. 132–133). However, rectus femoris also acts over the hip joint, and is able to assist hip flexion as well.

Attachments

- There are two proximal attachment points:
  - anterior inferior iliac spine (straight head);
  - notch above the acetabulum and capsule of the hip joint (reflected head).
- Common distal quadriceps attachment to the ligamentum patellae and tibial tuberosity via the patella.

Nerve supply

Femoral nerve (L2, L3, L4).

Functions

- ♦ Assists flexion of the hip.
- ♦ Extension of the knee.

Study tasks

- Shade lines between the attachment points and consider the muscle functions.
- Test the muscle on a colleague lying supine with both hip and knee flexed to 90°. Resist their continuing hip flexion.

Sartorius ('tailor's muscle')

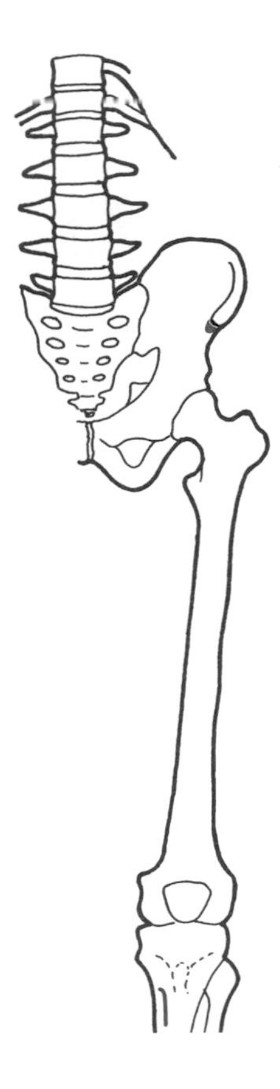

Sartorius is elegant in both name and function. It is one of the best examples of a 'strap' muscle, which sacrifices power for an impressive range of movement. It crosses both hip and knee joints obliquely, and is therefore able to flex both joints and produce rotation at the same time. It enables the leg to be crossed so that the ankle rests on the contralateral knee in the manner favoured by tailors whilst sewing, hence the name.

Attachments

- Anterior superior iliac spine.
- Medial surface of the upper tibia.

Nerve supply

Femoral nerve (L2, L3).

Functions

- Assists flexion of the hip.
- Assists flexion of the knee.
- Assists abduction with lateral rotation of the hip.
- Assists medial rotation of the knee.

Study tasks

- Shade lines between the attachment points, and consider the muscle functions.
- Test the muscle on a supine colleague. Their hip and knee should be flexed to 90°, with the hip in slight abduction and lateral rotation. Hold their heel and resist their attempt to place the heel on their contralateral knee.

Gracilis ('slender muscle')

Like sartorius, this is a 'strap' muscle which shows similar versatility, but its attachments give a rather more direct line of pull.

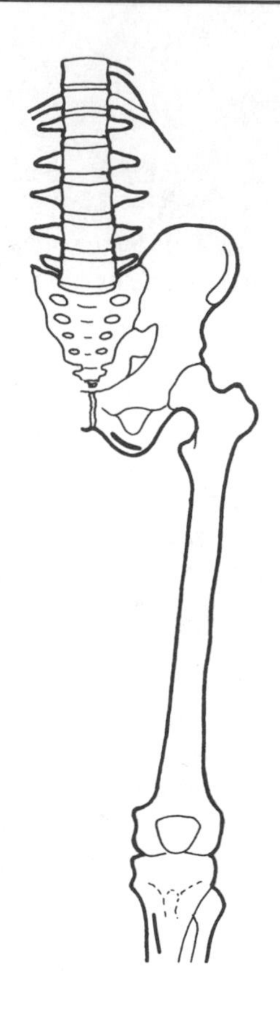

Attachments

- Medial margin of the inferior pubic ramus.
- Medial surface of the upper tibia.

Nerve supply

Obturator nerve (L2, L3).

Functions

- Assists adduction of the hip.
- Assists medial rotation of the hip.
- Assists medial rotation of the knee.

Study tasks

- Shade lines between the attachment points and consider the muscle functions.
- Test the muscle on a supine colleague with their leg held straight and in medial rotation with the arch of the foot resting on the contralateral foot. Try to push their leg away into abduction while they resist.

Extension

It should be noted that the hamstring muscles flex the knee as well as extend the hip. With the knee extended the hip has a greater range (40°) than when the knee is flexed (30°). This is due to the fact that knee flexion contracts the hamstrings, leaving less power available for hip extension. During extension all the hip ligaments tighten, especially the iliofemoral ligament.

Note also that extension of the hip may be functionally increased by anterior tilt of the pelvis, thus exaggerating the lumbar lordosis.

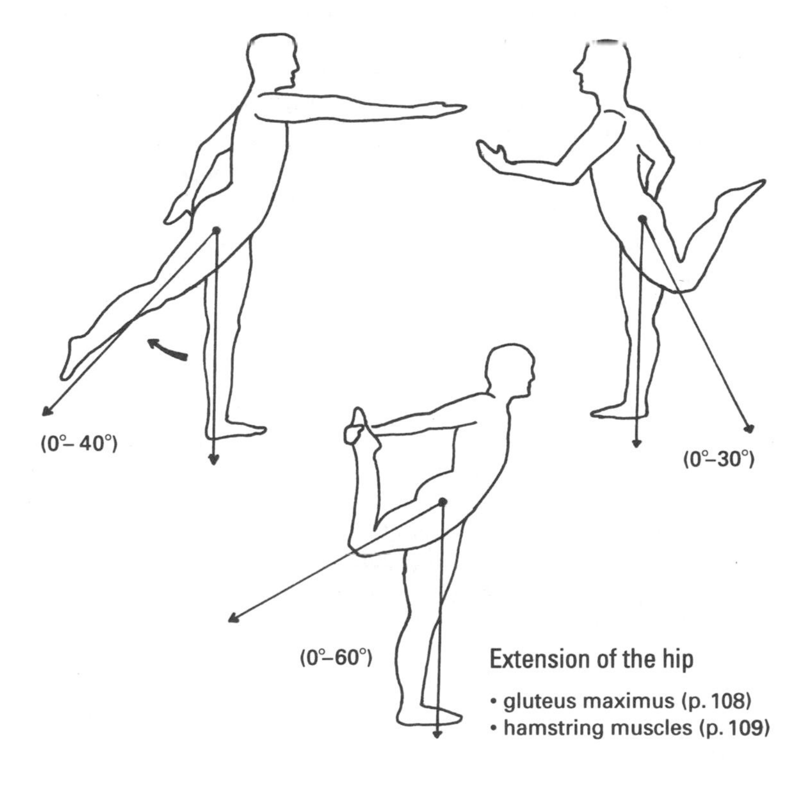

Extension of the hip

- gluteus maximus (p. 108)
- hamstring muscles (p. 109)

Study tasks

- Highlight the names of the muscles.
- Study the details of each muscle with reference to the page number given.

Gluteus maximus ('largest muscle of the buttocks)

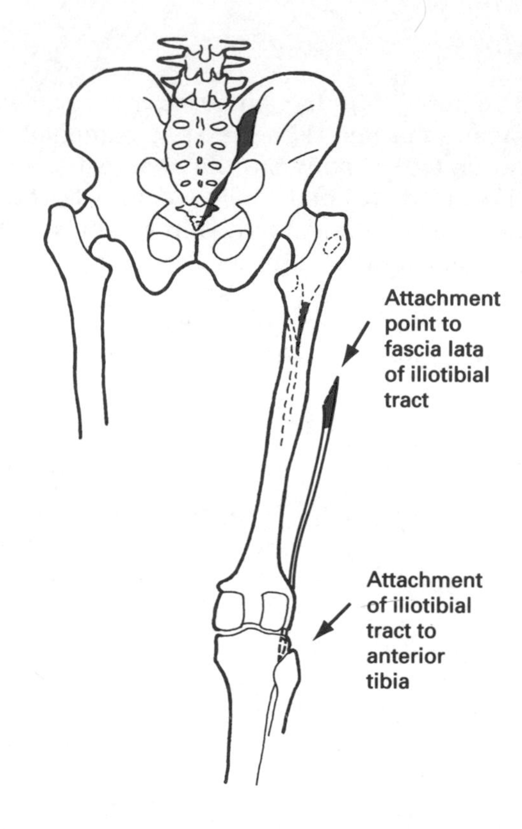

Attachments

- Posterior to the posterior gluteal line on the ilium.
- Posterior surfaces of the sacrum and coccyx.
- The sacrotuberous ligament (not shown).
- The lumbar fascia (not shown).
- The gluteal tuberosity of the femur.
- The fascia lata of the iliotibial tract.

Nerve supply

Inferior gluteal nerve (L5, S1, S2).

Functions

- Extension of the hip, especially from the flexed position, as in climbing the stairs.
- Assists lateral rotation of the hip.
- Assists abduction by tightening the iliotibial tract.
- Opposes anterior tilt of the pelvis, and thereby helps to maintain good posture.

Study tasks

- Shade lines from the first two attachment points, so that they converge on the gluteal tuberosity and fascia lata, and consider the various muscle functions.
- Test the muscle on a prone colleague with their knee flexed to 90° and a pillow under their thigh, to give slight hip extension. Press firmly over the posterior thigh as they try to extend the hip further.

The hamstring group ('tendons at the back of the knee')

(i) Biceps femoris ('two-headed muscle of the thigh')

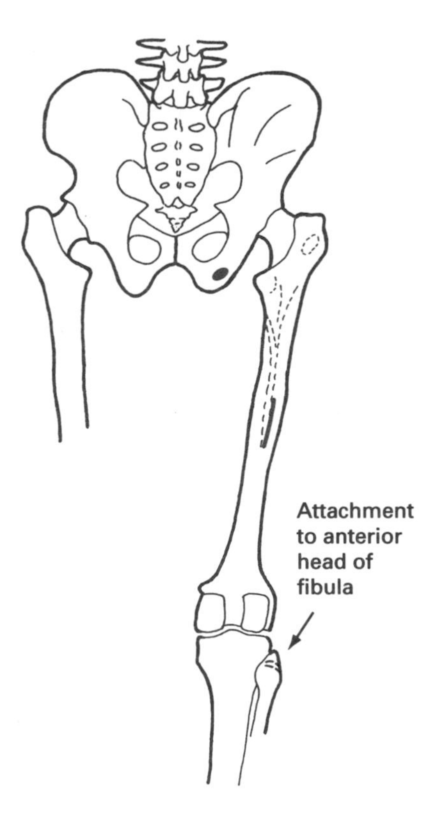

Attachments

- Medial aspect of the ischial tuberosity (long head).
- Linea aspera (short head).
- Head of the fibula (surrounding the collateral ligament).

Nerve supply

Sciatic nerve (L5, S1, S2).

Functions

- Extension of the hip.
- Flexion of the knee.
- Lateral rotation of the flexed knee.

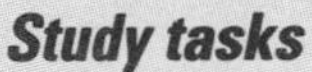

Study tasks

- Shade lines between the attachment points and consider the muscle functions.
- Palpate the muscle tendon on yourself in a sitting position. It is easily felt on the postero-lateral aspect of the thigh running towards the head of the fibula.
- Test the muscle by asking a colleague to lie prone with their knee flexed to 90°. Laterally rotate their thigh so that the surface of the muscle faces up. Resist with your hand as they try to pull heel to buttock.

(ii) Semitendinosus ('semi-tendinous muscle')

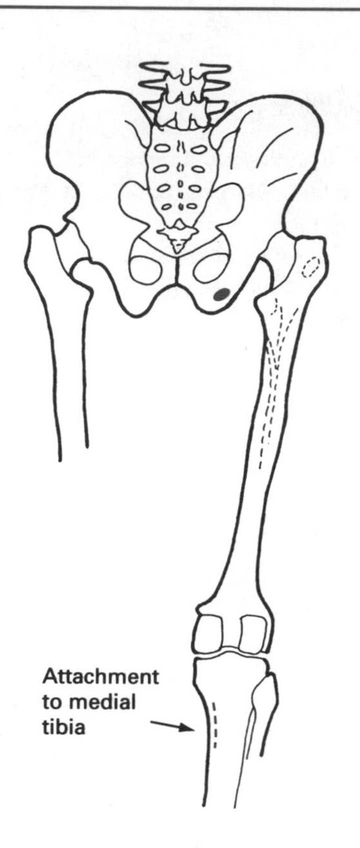

Attachments

- Medial aspect of the ischial tuberosity.
- Medial aspect of the superior surface of the tibia.

Nerve supply

Sciatic nerve (L5, S1, S2).

Functions

- Extension of the hip.
- Flexion of the knee.
- Medial rotation of the flexed knee.

Study tasks

- Shade lines between the attachment points and consider the muscle functions.
- Palpate the tendinous nature of the muscle on yourself in a sitting position. It is the largest and most lateral of the palpable tendons on the medial aspect of the posterior thigh.
- Test the muscle by asking a colleague to lie prone with their knee flexed to 90°. Medially rotate their thigh so that the muscle faces up. Resist with your hand as they try to pull heel to buttock.

Attachments

- Lateral aspect of the ischial tuberosity.
- Posterior surface of the medial tibial condyle and oblique popliteal ligament and fascia.

Nerve supply

Sciatic nerve (L5, S1, S2).

Functions

- Extension of the hip.
- Flexion of the knee.
- Medial rotation of the flexed knee.

Study tasks

- Shade lines between the attachment points and consider the muscle functions.
- Palpate the membranous nature of the tendon on yourself in the sitting position. It may be felt medial to the large semitendinosus tendon. However the smaller tendon of gracilis is also palpable here. Semimembranosus lies medial to this.
- Test the muscle as for semitendinosus above.

(iii) Semimembranosus ('semi-membranous muscle')

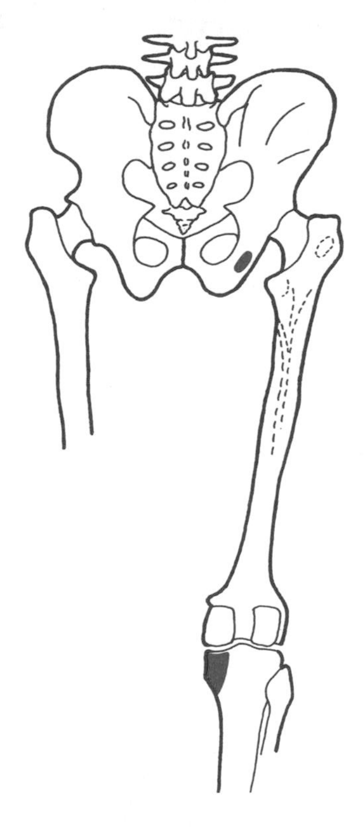

Abduction

In the standing position, abduction of one hip from the midline results in the recruitment of the contralateral abductor muscles as well, in order to support the body. This highlights the routine use of the abductor muscles in supporting the body whilst walking, which requires momentarily standing on one leg.

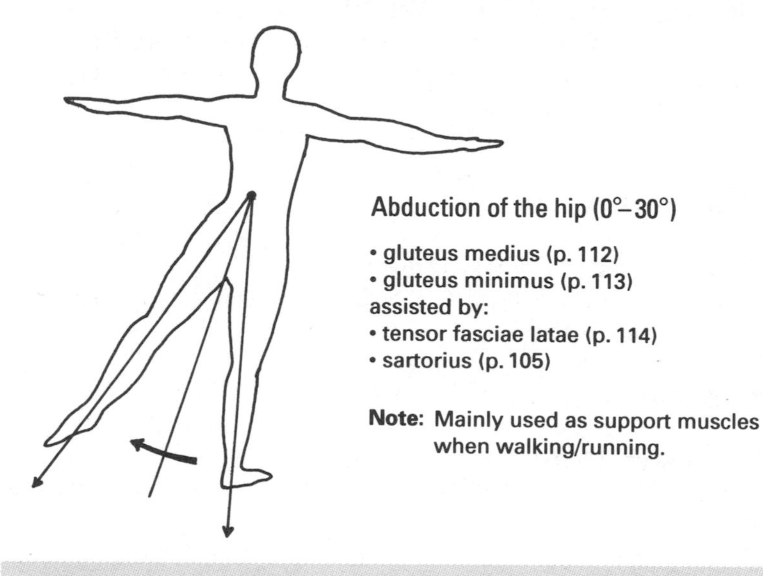

Abduction of the hip (0°–30°)

- gluteus medius (p. 112)
- gluteus minimus (p. 113)

assisted by:

- tensor fasciae latae (p. 114)
- sartorius (p. 105)

Note: Mainly used as support muscles when walking/running.

Study tasks

- Highlight the names of the muscles.
- Study the details of each muscle with reference to the page number given.

Gluteus medius ('middle-sized muscle of the buttock')

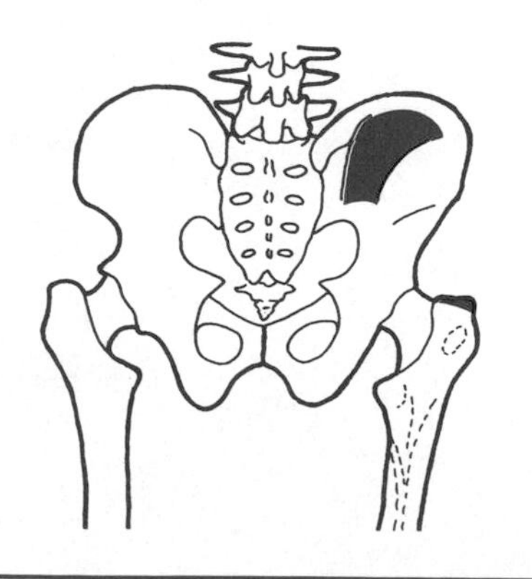

Attachments

- Between the posterior and anterior gluteal lines on the ilium.
- Greater trochanter of the femur.

Nerve supply

Superior gluteal nerve (L5, S1).

Functions

- Abduction of the hip.
- Medial rotation of the hip (anterior fibres).
- Supports the body in the 'mid-stance' phase of walking/running.

Study tasks

- Shade lines between the attachment points and consider the muscle functions.
- How might the concept of 'origin' and 'insertion' vary with the different usage of this muscle?
- Test this muscle with gluteus minimus (below) by asking a colleague to lie supine with their leg in slight abduction. Resist on the lateral side of their leg, as they try to abduct further.

Gluteus minimus ('smallest muscle of the buttock')

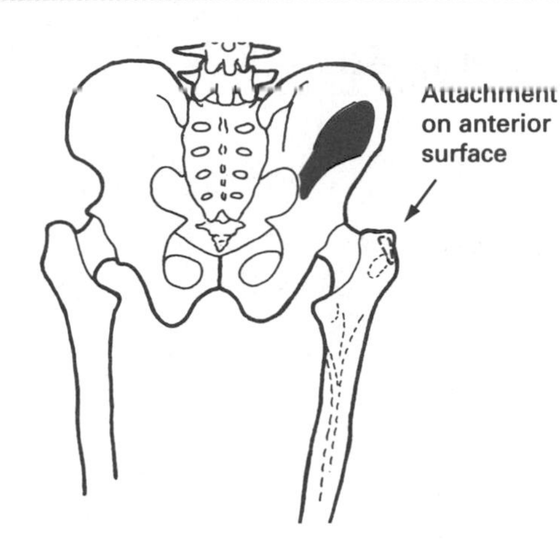

Attachments

- Between the inferior and anterior gluteal lines on the ilium.
- Anterior part of the greater trochanter.

Nerve supply

Superior gluteal nerve (L5, S1).

Functions

- Abduction of the hip.
- Medial rotation of the hip (anterior fibres).
- Supports the body in the 'mid-stance' phase of walking/running.

Study tasks

- Shade lines between the attachment points and consider the muscle functions.
- How might the concept of 'origin' and 'insertion' vary with the different usage of this muscle?
- Test the muscle with gluteus medius (p. 112).

Tensor fasciae latae ('the muscle which tightens the wide fascia')

Iliotibial tract

This muscle tightens the thickened wide fascia (fascia lata) on the lateral part of the thigh which forms the iliotibial tract. It acts from a distance, and is perhaps best regarded as an assistor in most of its functions.

Attachments

- Anterior part of the iliac crest and anterior superior iliac spine.
- Iliotibial tract approximately 7cm below the greater trochanter.

Nerve supply

Superior gluteal nerve (L5, S1).

Functions

- ♦ Assists abduction of the hip.
- ♦ Medial rotation of the hip.
- ♦ Assists knee extension, with some lateral rotation.

Study tasks

- Shade lines between the attachment points and consider the functions of the muscle and iliotibial tract.
- Test the muscle by asking a colleague to lie supine. Palpate the iliotibial tract with their knee alternately in flexion and extension. Test the hip functions with the thigh in slight abduction and medial rotation. Resist further abduction of the hip.

Adduction

Pure adduction from the anatomical position is not possible, because both limbs are in contact. Adduction is possible in the supine position if the contralateral leg is raised. However, relative adduction takes place from a position of abduction; and adduction may take place combined with flexion or extension, or when gripping with the knees is required.

Adduction of the hip (shown with extension) (0°–30°)

- adductor longus (p. 115)
- adductor brevis (p. 116)
- adductor magnus (p. 116)

assisted by:

- pectineus (p. 103)
- gracilis (pp. 105–106)

Study tasks

- Highlight the names of the muscles.
- Study the details of each muscle with reference to the page number given.

Adductor longus ('long adductor muscle')

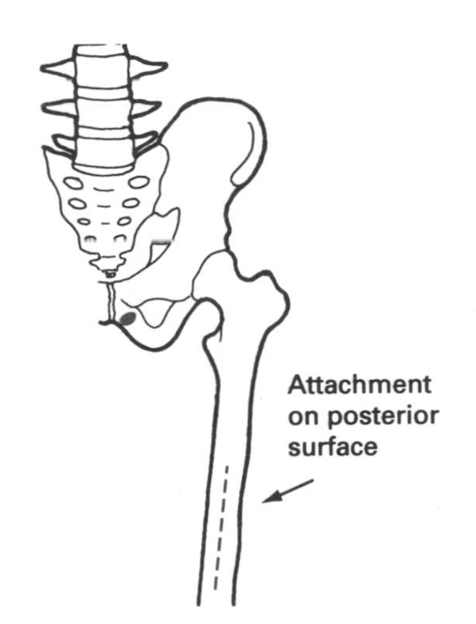

Attachments

- Pubis, below pubic crest.
- Middle part of the linea aspera on the femur.

Nerve supply

Obturator nerve (L2, L3, L4).

Functions

- Adduction of the hip.
- Assists lateral rotation of the hip.
- Assists flexion of the extended hip.

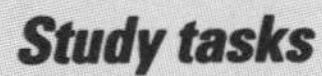

Study tasks

- Shade lines between the attachment points and consider the muscle functions.
- Test the muscle on a colleague lying supine with legs in slight abduction. Resist while one or both legs are adducted. It is difficult to differentiate between the adductor muscles, so use this as a test for all hip adductors.

Adductor brevis ('short adductor muscle')

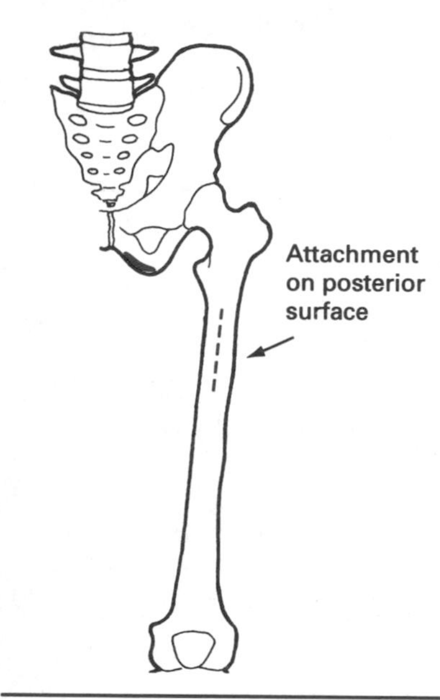

Attachments

- Inferior ramus and outer body of the pubis.
- Upper part of the linea aspera on the femur.

Nerve supply

Obturator nerve (L2, L3, L4).

Functions

- Adduction of the hip.
- Assists in flexion of the extended hip.

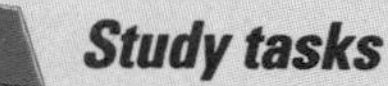

Study tasks

- Shade lines between the attachment points and consider the muscle functions.
- Test the muscle as for adductor longus (p. 115).

Adductor magnus ('large muscle of adduction')

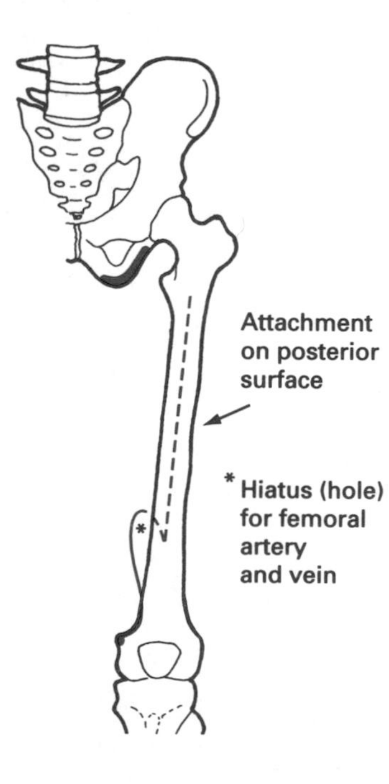

Attachments

- Ischial tuberosity and ramus; inferior pubic ramus.
- The medial supracondylar ridge and adductor tubercle on the femur, via the linea aspera.

Nerve supply

Obturator nerve and sciatic nerve (L2, L3, L4).

Functions

- Adduction of the hip.
- Assists flexion of the extended hip.
- Assists extension of the hip (its attachments enable it to act with the hamstrings as a hip extensor).

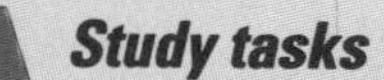

Study tasks

- Shade lines between the attachment points and consider the muscle functions.
- Test the muscle as for adductor longus above.

Medial rotation

In comparison with lateral rotation, medial rotation of the hip is relatively weak, and the range of movement is smaller.

Medial rotation of the hip (0°–30°)

- tensor fasciae latae (p. 114)
- gluteus medius and minimus (pp. 112–113) (anterior fibres)

assisted by:

- gracilis (pp. 105–106)
- adductor muscles (p. 114) (limited according to starting position)

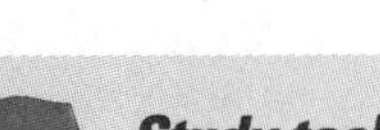

Study tasks

- Highlight the names of the muscles shown.
- Study the details of each muscle with reference to the page number given.
- The gluteal muscles are sometimes called 'the deltoid of the hip'. Would you consider this to be an appropriate description? (See also p. 32.)

Lateral rotation

Lateral rotation is more powerful than medial rotation, and has a greater range of movement.

Lateral rotation of the hip (0°–60°)

- obturator internus (p. 118)
- obturator externus (p. 118)
- gemellus superior (p. 119)
- gemellus inferior (p. 119)
- quadratus femoris (p. 120)

assisted by:

- piriformis (p. 121)
- gluteus maximus (p. 108)
- sartorius (p. 105)

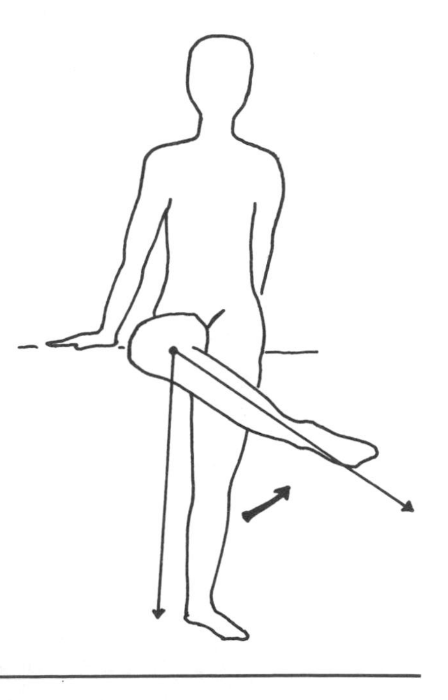

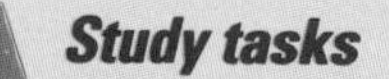

Study tasks

- Highlight the names of the muscles.
- Study the details of each muscle with reference to the page number given.

Obturator internus ('internal covering muscle')

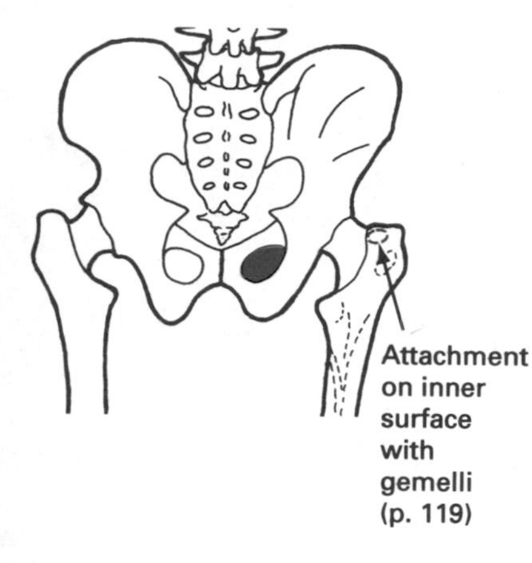

Attachments

- Inner surface of the obturator membrane and internal surface of the pelvis below the greater sciatic foramen.
- Internal surface of the greater trochanter of the femur.

Nerve supply

Nerve to obturator internus (L5, S1).

Functions

- Lateral rotation of the extended hip.
- Abduction of the flexed hip.

Study tasks

- (The anatomical details of this muscle are quite complex and only part of the attachments can be shown here. It may be helpful to consult *Gray's Anatomy* (1995) and refer to a skeletal specimen.)
- Shade lines between the attachment points and consider the functions of this muscle. The muscle bends in a pulley-like fashion across the groove between the ischial spine and tuberosity.
- Test the muscle on a colleague who should be placed prone with a pillow under the knee to induce slight hip extension. Resist by restraining their ankle while they rotate the thigh laterally, with knee flexed.

Obturator externus ('external covering muscle')

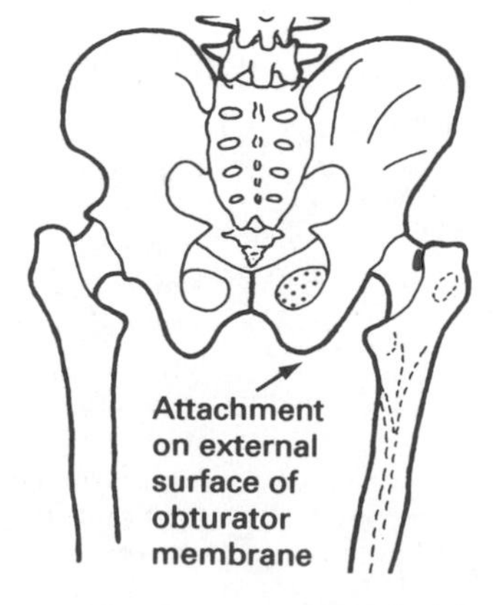

Attachments

- External surface of the obturator membrane and surrounding bone.
- Trochanteric fossa of the femur.

Nerve supply

Obturator nerve (L3, L4).

Function

- Lateral rotation of the hip.

Study tasks

- Shade lines between the attachment points and consider the muscle functions.
- Test the muscle as for obturator internus.

The gemelli ('the paired muscles')

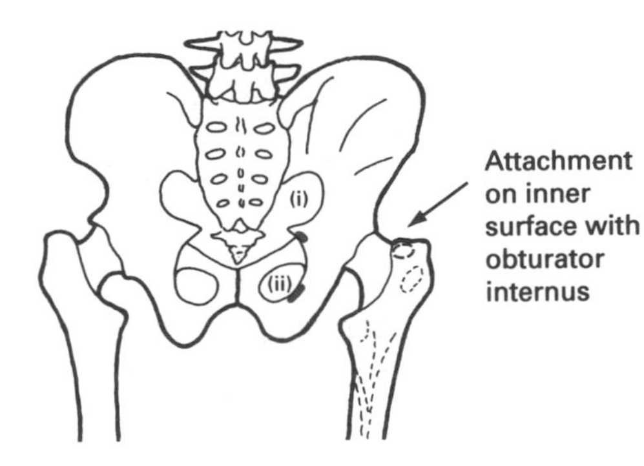

(i) Gemellus superior ('upper paired muscle')

Attachments

- Ischial spine (above the lesser sciatic notch).
- The medial surface of the greater trochanter of the femur (blending with the insertion of obturator internus).

Nerve supply

Nerve to obturator internus (L5, S1).

(ii) Gemellus inferior ('lower paired muscle')

Attachments

- Upper part of the ischial tuberosity.
- The medial surface of the greater trochanter of the femur (blending with the insertion of obturator internus).

Nerve supply

Nerve to quadratus femoris (L5, S1).

Functions

♦ Lateral rotation of the extended hip.
♦ Abduction of the flexed hip.

Study tasks

- Shade lines between the attachment points of the gemelli so that they converge onto the greater trochanter, and consider their functions.
- Test the muscles as for obturator internus.

Quadratus femoris ('square-shaped muscle of the thigh')

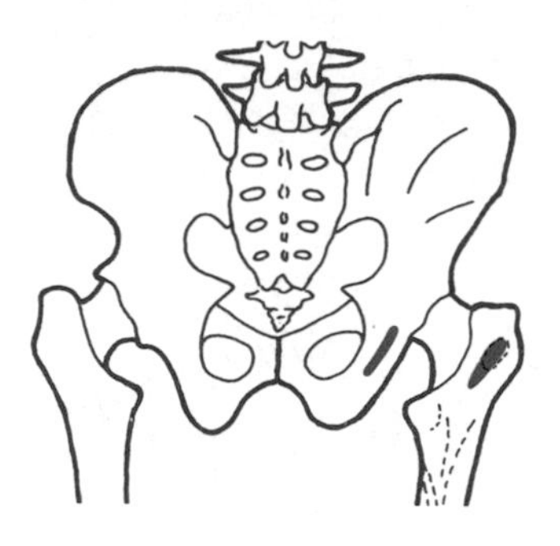

Attachments

■ Lateral border of the ischial tuberosity.
■ Quadrate tubercle on posterior surface of the greater trochanter.

Nerve supply

Nerve to quadratus femoris (L5, S1).

Function

♦ Lateral rotation of the hip.

Study tasks

- Shade lines between the attachment points and consider the muscle function.
- Test the muscle by placing a colleague in the prone position with one knee flexed. Resist the movement of their ankle while they try to rotate their thigh laterally.

Piriformis ('pear-shaped muscle')

Anterior view

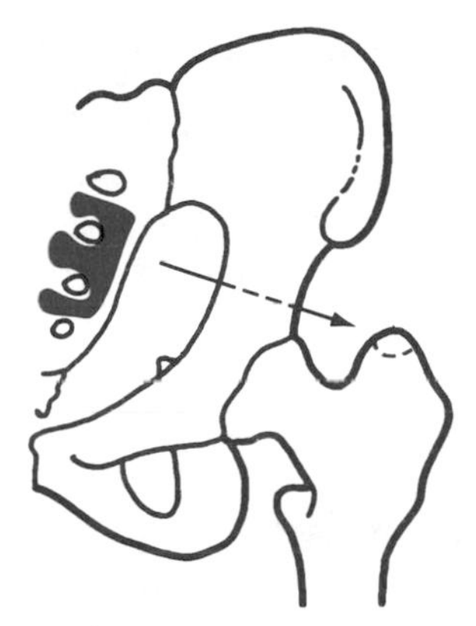

Posterior view

Attachments

- Pelvic surface of the sacrum.
- Superior surface of the greater trochanter of the femur, having passed through the greater sciatic foramen.

Nerve supply

(L5, S1, S2).

Functions

- Lateral rotation of the extended hip.
- Abduction of the flexed hip.

Study tasks

- Shade lines between the attachment points and consider the muscle functions.
- Test this muscle as for obturator internus.
- Consider the movement of circumduction with reference to the hip and the muscles producing the movement.

The muscles of the pelvic diaphragm

Unlike the thoracic diaphragm (p. 168), the pelvic diaphragm does not consist of a single muscle, but an amalgamation of a number of muscles which form a supporting mechanism. They support the pelvic viscera from below, rather like a pair of clasped hands. This is important when intra-abdominal pressure is increased from above by the descent of the thoracic diaphragm. The anatomical details are complex, and a standard anatomy text should be consulted if greater detail is required.

Levator ani ('muscle which lifts the anal region')

(Pelvis: superior view)

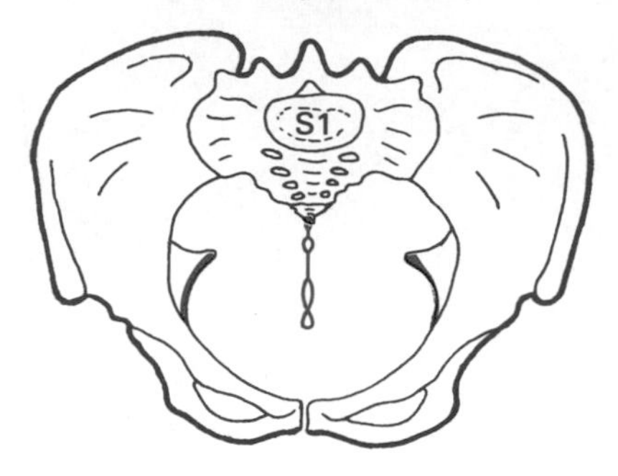

Attachments

- The inner surface of the lower pelvis on each side.
- A raphe in the midline (median plane) which can be subdivided into:
  posterior fibres inserting into the tip of the coccyx;
  middle fibres surrounding the rectum and anal canal;
  anterior fibres surrounding the prostate or vagina which either elevate the prostate in the male, or act as a vaginal sphincter in the female.

Nerve supply

S4 and the perineal branch of the pudendal nerve, or inferior rectal nerve.

Functions

- Sphincter actions at the anorectal junction and of the vagina.
- Maintains the position of the pelvic organs.
- Acts with the thoracic diaphragm and abdominal muscles to maintain intra-abdominal pressure during exertion.

Study tasks

- Shade lines between the attachment points and consider the function of the muscle as a supporting mechanism.
- The muscle can be self-tested by squeezing gently but firmly with the anal muscles.
- Consider the muscular action of the Valsalva manoeuvre (p. 169).

Coccygeus ('tail muscle')

This small muscle is posterior to, and continuous with, levator ani.

(Pelvis: superior view)

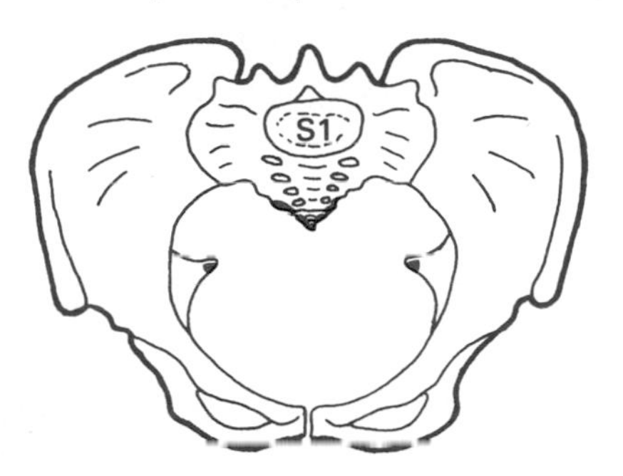

Attachments

- Ischial spine and sacrospinous ligament.
- Anterolateral border of the coccyx and lower sacrum.

Nerve supply

Branch from S4, S5.

Functions

- Flexion of the coccyx.
- Acts with levator ani to support the pelvic viscera.

Study task

- Shade lines between the attachment points and consider the muscle functions.

Chapter 9

The knee

Introduction

In the anatomical position, the knee is already in extension. The angle of inclination seen at the hip (p. 98) means that the long axis of the femoral shaft cannot coincide with that of the tibia. The shaft of the tibia therefore forms an obtuse angle of approximately 170°-175° with the shaft of the femur. This effectively brings the knees together, and may be referred to as the 'normal' or 'natural valgus' of the knee.

The natural valgus of the knee

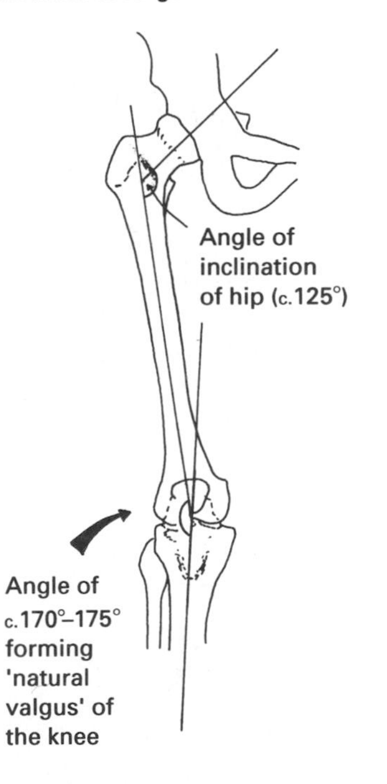

In the same way that the elbow provides the upper limb with a hinge, so the knee allows the lower limb the flexibility of a versatile hinge. However, there is a price to be paid. The knee, unlike the elbow, is weight-bearing, and the amount of rotation available is strictly limited. The main stresses of movement and shock absorption on weight-bearing must be borne by the hingeing mechanism and the powerful extensor muscles. The condylar nature of the hinge joint allows some rotation, but often the vulnerability of the hinge is exposed, particularly in sport.

The bones and joints

The knee consists of the articulations between the distal end of the femur; the proximal surface of the tibia; the fibula; and the patella, which is a sesamoid bone.

The highly specialized shape of the distal surface of the femur articulates with the proximal surface of the tibia. The medial femoral condyle is more curved than the lateral condyle, and this allows a limited degree of rotation when the knee is moving from flexion to extension, and vice versa. The congruity of the joint (fit) is increased by the presence of the fibrocartilaginous menisci.

The bones of the knee

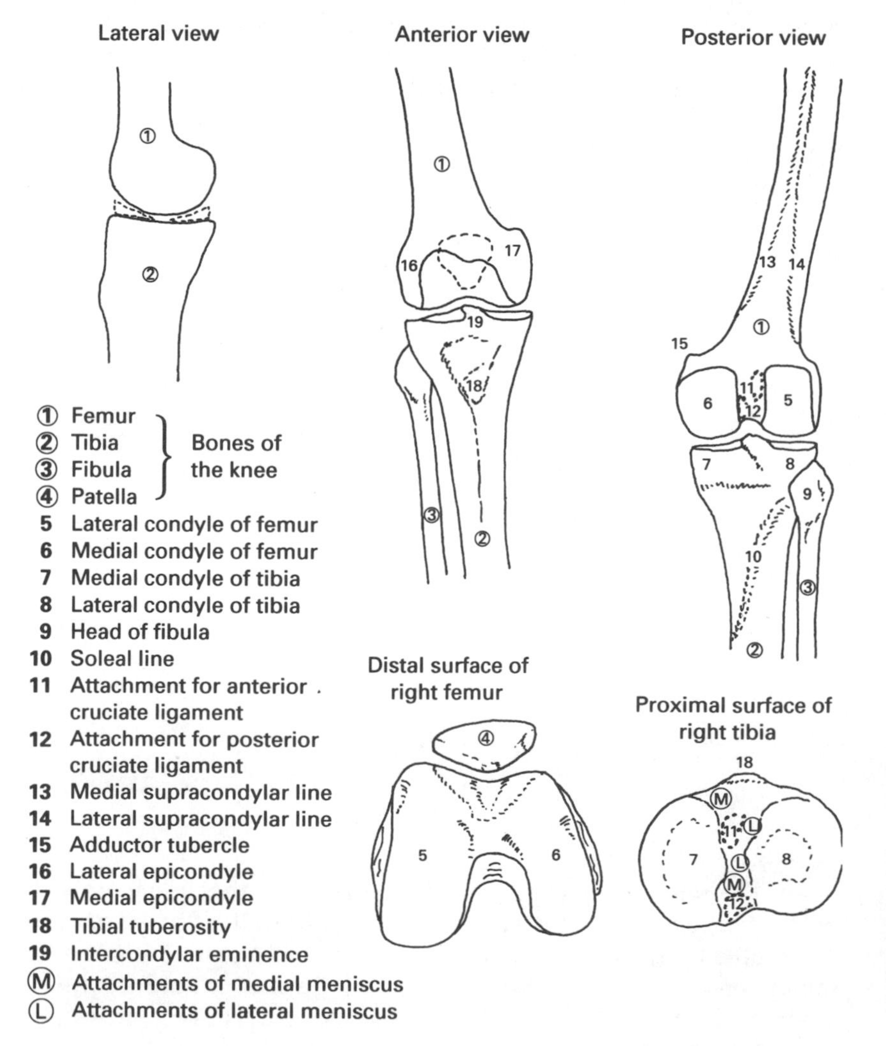

Study tasks

- Colour the bones in separate colours.
- Highlight the names of the bone features.

Joints and ligaments of the knee (anterior view)

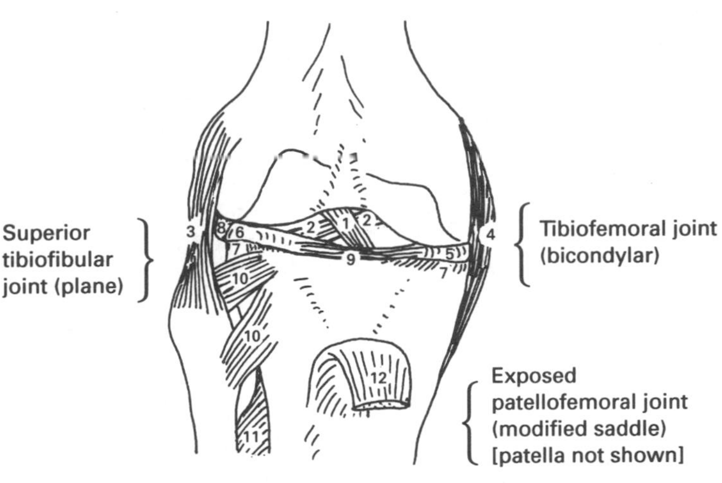

1. Anterior cruciate ligament
2. Posterior cruciate ligament
3. Lateral (fibular) collateral ligament
4. Medial (tibial) collateral ligament
5. Medial meniscus
6. Lateral meniscus
7. Coronary ligaments
8. Popliteus tendon
9. Transverse ligament
10. Anterior ligament of head of fibula
11. Interosseous membrane
12. Ligamentum patellae (cut, and turned inferiorly)

Study tasks

- Highlight the names of the joints and study the details given.
- Shade the ligaments and highlight the features as indicated.
- Palpate the joint line between the tibia and femur on yourself or a colleague. It can be felt most readily at the sides, and at the level of the inferior pole of the patella, in a relaxed, slightly flexed knee.
- Palpate the collateral ligaments. The medial collateral ligament may be felt as a resistant band over the joint line medially. The lateral collateral ligament is palpable if a finger is placed in the hollow just above the head of the fibula, and the knee placed in a cross-legged position. If the thigh is laterally rotated the ligament can be felt as it tightens. Do not confuse with the tendon of biceps femoris muscle which also inserts into the head of the fibula (p. 109).

The attachments of the capsule and synovial membrane are complex. The fibrous capsule is attached to the margins of the articular surfaces, and surrounds the joint except anteriorly. Here, it allows the synovial membrane which lines the capsule to pouch upwards to form the suprapatellar bursa. The synovial membrane is attached to the tibial surface in a way that allows the menisci to lie within, but the cruciate ligaments to lie outside, its walls.

Superior articular surface of right tibia (oblique view)

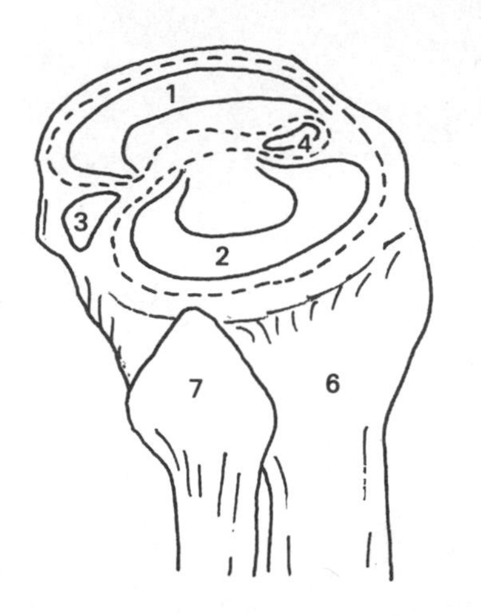

1 Medial meniscus
2 Lateral meniscus
3 Attachment for posterior cruciate ligament
4 Attachment for anterior cruciate ligament
5 Femur
6 Tibia
7 Fibula
8 Patella
9 Quadriceps tendon
10 Prepatellar bursa
11 Ligamentum patellae
12 Deep infrapatellar bursa
13 Superficial infrapatellar bursa
14 Articularis genus muscle*
15 Infrapatellar fat pad
16 Suprapatellar bursa
- - - Synovial membrane

*A small muscle which suspends the suprapatellar bursa and attaches to the anterior shaft of the femur

Knee
(sagittal view)

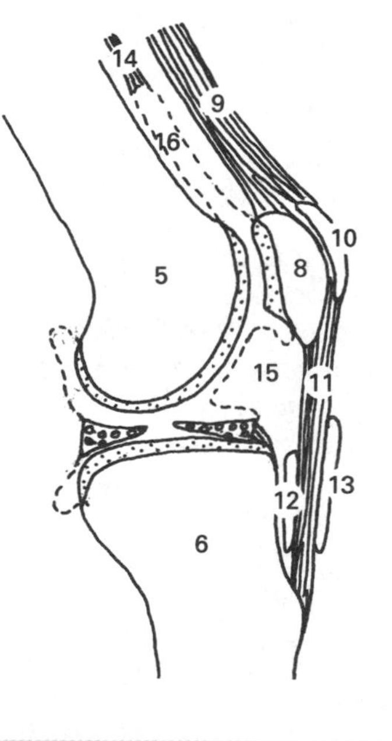

Study tasks

- Highlight the names of the main features shown.
- Colour the outline of the synovial membrane.

The bursae of the knee joint are numerous, and only the main ones are shown here.

The muscles

Identification of the superficial muscles and tendons surrounding the knee is not always easy, but can be improved with palpation and practice.

The superficial muscles

Anterior view

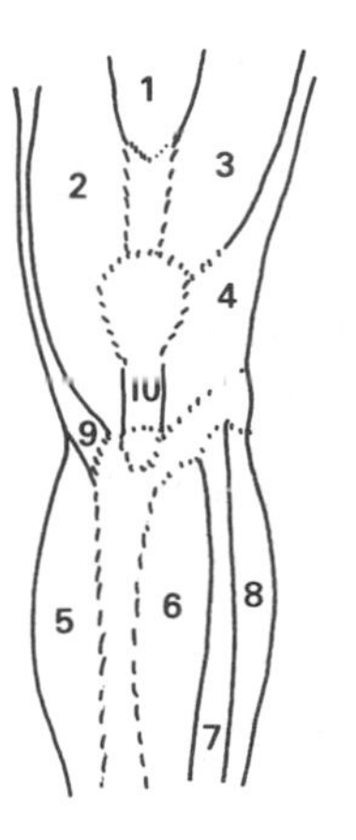

1 Rectus femoris (p. 134)
2 Vastus medialis (p. 135)
3 Vastus lateralis (p. 137)
4 Iliotibial tract
5 Gastrocnemius (p. 147)
6 Tibialis anterior (p. 144)
7 Extensor digitorum longus (p. 145)
8 Peroneus longus (p. 156)
9 Sartorius (p. 105)
10 Ligamentum patellae

Posterior view

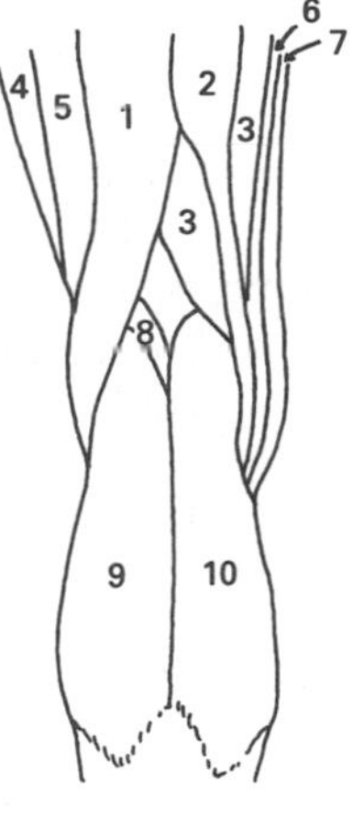

1 Biceps femoris (p. 109)
2 Semitendinosus (p. 110)
3 Semimembranosus (p. 111)
4 Iliotibial tract
5 Vastus lateralis (p. 137)
6 Gracilis (pp. 105–106)
7 Sartorius (p. 105)
8 Plantaris (p. 149)
9 Gastrocnemius (lateral head) (p. 147)
10 Gastrocnemius (medial head) (p. 147)

The superficial muscles (flexed knee)

Medial view

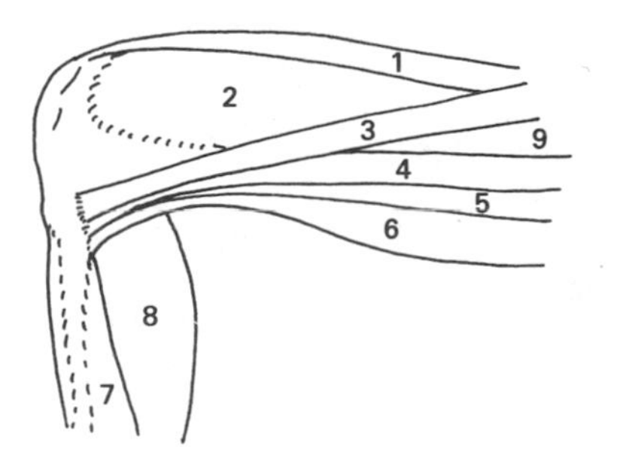

1 Rectus femoris (p. 134)
2 Vastus medialis (p. 135)
3 Sartorius (p. 105)
4 Gracilis (pp. 105–106)
5 Semimembranosus (p. 111)
6 Semitendinosus (p. 110)
7 Soleus (p. 148)
8 Gastrocnemius (p. 147)
9 Adductor longus (p. 115)

Lateral view

1 Rectus femoris (p. 134)
2 Vastus lateralis (p. 137)
3 Iliotibial tract
4 Biceps femoris (p. 109)
5 Gastrocnemius (p. 147)
6 Peroneus longus (p. 156)
7 Extensor digitorum longus (p. 145)
8 Tibialis anterior (p. 144)

Study tasks

- Shade the muscles shown in separate colours.
- Highlight the names of the muscles.

Movements

The knee travels principally from extension in the anatomical position, to flexion which brings the foot closer to the thigh. The condylar nature of the joint permits a measure of rotation which offers manoeuvrability when weight-bearing, and also allows a degree of final 'lock' as the femur rotates medially on the tibia in extension on weight-bearing.

The bicondylar nature of the tibiofemoral joint allows a greater degree of movement than is normally seen in a hinge joint.

Flexion and extension produce a rolling/sliding/gliding movement of the femoral condyles on the surface of the tibia. The integrity of this movement is controlled by the deformable nature of the fibrocartilaginous menisci which follow the movement of the femoral condyles, posteriorly in flexion and anteriorly in extension.

Movements of the menisci on the tibia during various movements of the knee

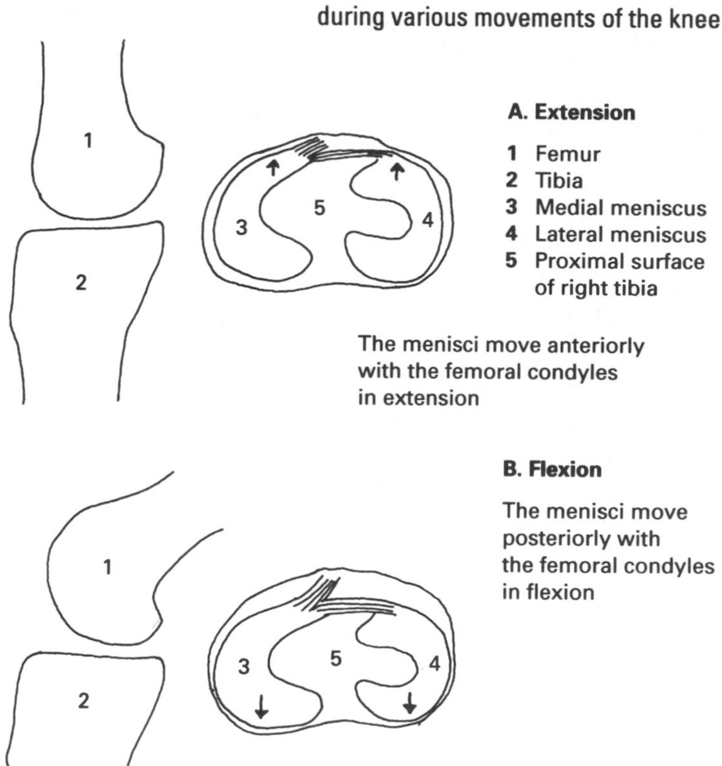

Study tasks

- Shade the menisci in colour.
- Study and highlight the details of the movements given.

Rotation occurs through a longitudinal axis, with the knee flexed or semiflexed. It occurs largely because of the wheel-like and curved nature of the medial femoral condyle. The cruciate ligaments tighten as the knee extends, and laterally rotate the tibia as it reaches full extension.

Rotatory movements also occur when the knee is semiflexed, especially during weight-bearing activities, and allow adaptability of the knee joint.

During all rotation movements the menisci follow the movements of the femoral condyle. For example, if the body is rotated to the right on the right knee, the femur will rotate laterally on the tibia. The lateral meniscus will move posteriorly while the medial meniscus will move anteriorly on the surface of the right tibia. Unfortunately, sudden changes of direction on weight-bearing can cause tears in the menisci if they fail to synchronize with the condylar movements. This can also happen in pure flexion/extension movements if performed abruptly.

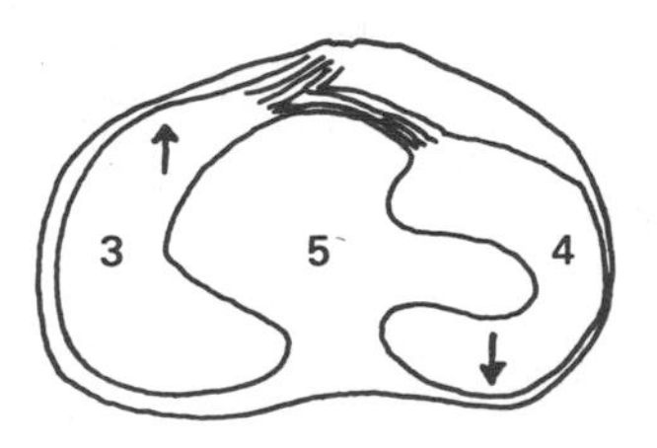

C. Medial rotation (of the leg and knee)

The medial meniscus moves anteriorly with the medial femoral condyle; the lateral meniscus moves posteriorly

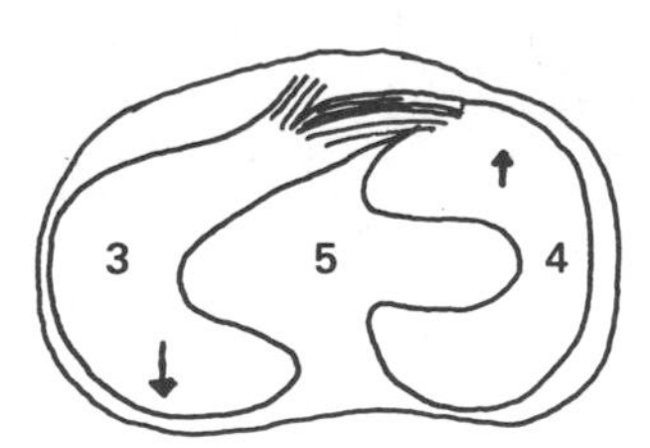

D. Lateral rotation (of the leg and knee)

The medial meniscus moves posteriorly with the medial femoral condyle; the lateral meniscus moves anteriorly

Flexion

In the foetal position the hips are laterally rotated and flexed in such a way that the thigh and calf face anteriorly, in a similar fashion to the upper limb.

Flexion therefore brings the calf and thigh closer together. It attains a greater range if the hip is flexed, since the hamstring muscles which produce the movement are also hip extensors (p. 107). Their action is shared between hip and knee so that more extension of the hip allows less flexion of the knee; while more flexion of the knee allows less extension of the hip.

Flexion of the knee

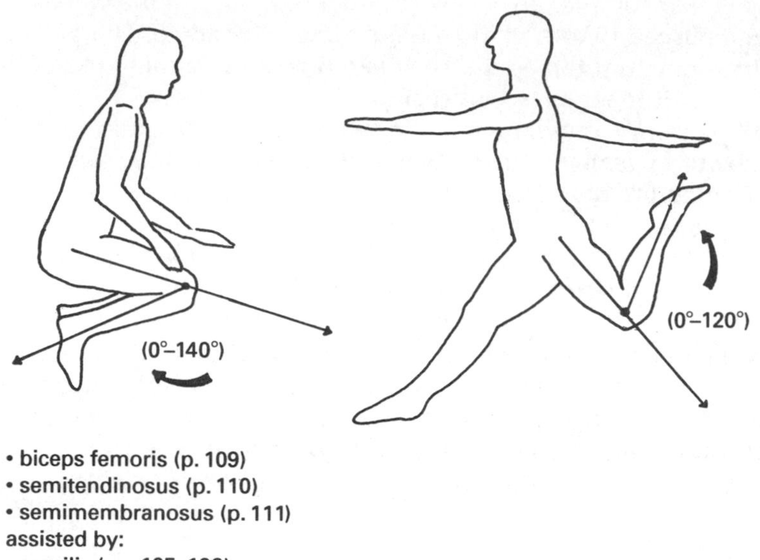

- biceps femoris (p. 109)
- semitendinosus (p. 110)
- semimembranosus (p. 111)

assisted by:

- gracilis (pp. 105–106)
- sartorius (p. 105)
- popliteus (p. 138)
- gastrocnemius (p. 147)
- plantaris (p. 149)

} when the foot is 'grounded'

Study tasks

- Demonstrate the varying degrees of knee flexion either on yourself or a colleague.
- Highlight the names of the muscles shown.
- Study the details of each muscle with reference to the page number given.

Extension

The knee is fully extended actively when the leg is held straight. Thus extension usually takes place from a position of relative flexion as in walking and running. It may demand considerable bursts of power such as climbing and kicking.

Extension of the knee

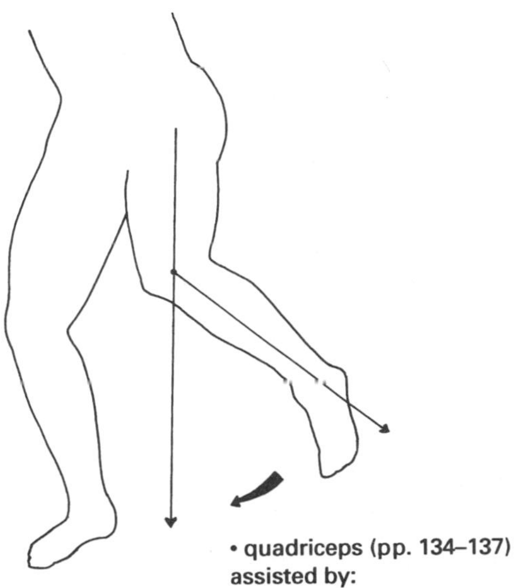

c.5° 'hyperextension'
available on
passive stretch

Study tasks

- Highlight the names of the muscles.
- Study the details of each muscle with reference to the page number given.

Quadriceps femoris ('four - headed muscle of the thigh')

Quadriceps femoris (colloquially the 'quads') is often referred to as one muscle, but is in fact four; and rectus femoris assists in hip flexion.

(i) Rectus femoris ('thigh-raising muscle')

Attachments

- There are two proximal attachment points:
  anterior inferior iliac spine (straight head);
  notch above the acetabulum and capsule of the hip joint (reflected head).
- Common distal quadriceps attachment to the ligamentum patellae and tibial tuberosity via the patella.

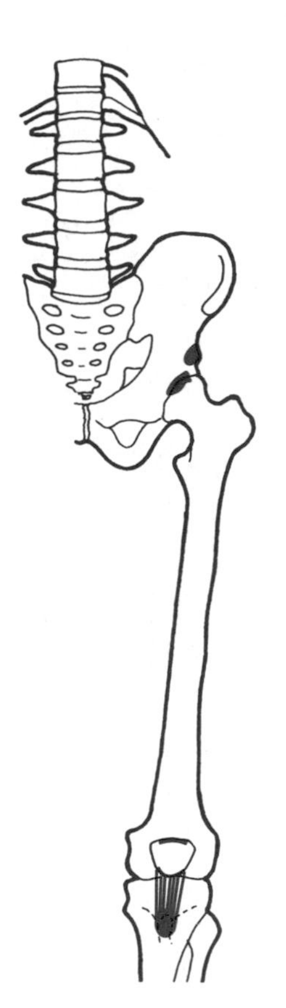

(ii) Vastus medialis ('very large inner muscle')

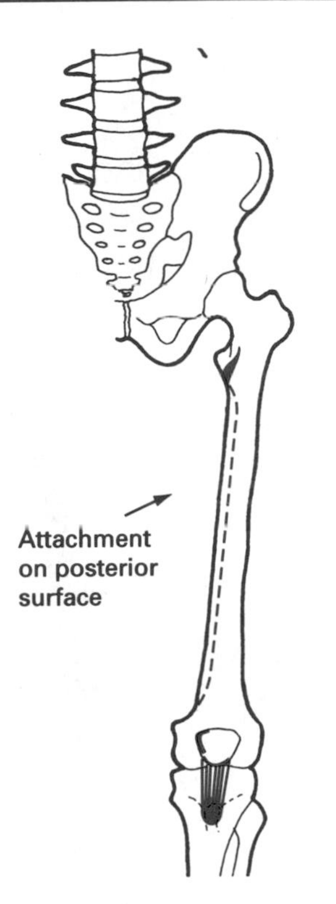

Attachments

- Intertrochanteric line; spiral line; medial linea aspera; medial supracondylar line of the femur.
- Adductor magnus and longus tendon (not shown).
- Common distal quadriceps attachment to the ligamentum patellae and tibial tuberosity via the patella.

(iii) Vastus intermedius ('very large middle muscle')

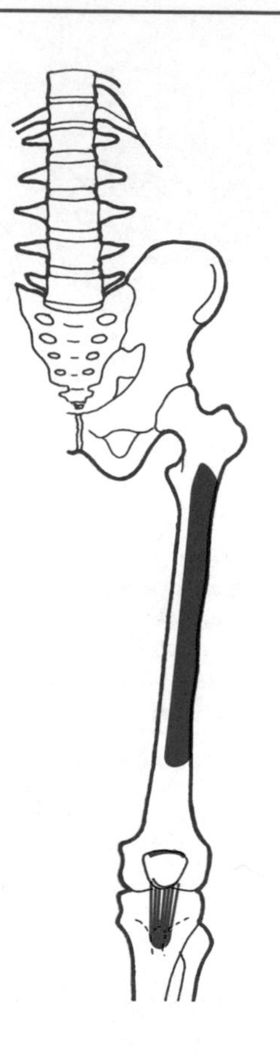

Attachments

- Anterolateral surface of the upper shaft of the femur.
- Common distal quadriceps attachment to the ligamentum patellae and tibial tuberosity via the patella.

Attachments

- Upper intertrochanteric line; anterior and inferior greater trochanter; lateral gluteal tuberosity; lateral linea aspera of the femur.
- Common distal quadriceps attachment to the ligamentum patellae and tibial tuberosity via the patella.

Nerve supply (all quadriceps)

Femoral nerve (L2, L3, L4).

Functions (all quadriceps)

- Extension of the knee.
- Rectus femoris alone assists flexion of the hip.

Study tasks

- Shade lines between the attachment points of the individual quadriceps muscles, and consider their functions.
- Test the quadriceps on a colleague lying supine with hip and knee slightly flexed. The knee being tested can be placed on the tester's arm which then rests on the colleague's opposite leg.
  Vastus medialis is tested by slightly laterally rotating the hip and then restraining extension of the knee.
  Vastus intermedius is tested by restraining extension of the knee without rotation at the hip.
  Vastus lateralis is tested by slightly medially rotating the hip and then restraining extension of the knee.
  Rectus femoris should be tested to bring out its functions as a hip flexor. Flex hip to 90° and restrain hip flexion by preventing the thigh being flexed towards the shoulder.
- Why is rectus femoris more effective as a knee extensor if the hip is extended?
- Consider how the function of the attachment points vary between the actions of kicking and rising out of an armchair.

(iv) Vastus lateralis ('very large outer muscle')

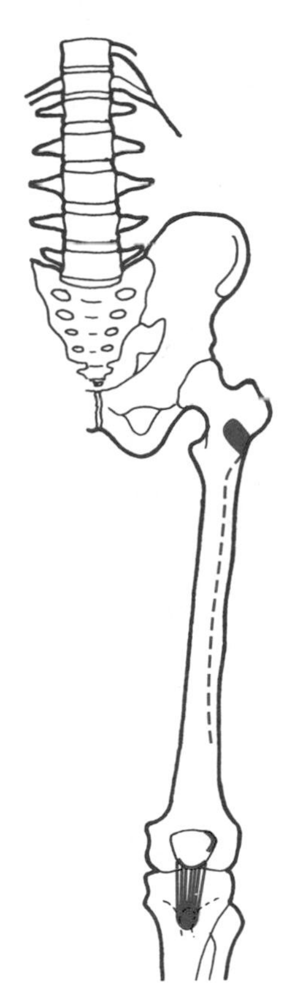

Rotation

Knee rotation movements refer to the rotation of the tibia on the femur. This occurs automatically as the tibia rotates slightly laterally from flexion to extension, and vice versa. Active rotation can also take place on a flexed or semiflexed knee.

Medial rotation of the knee (0°–30°)

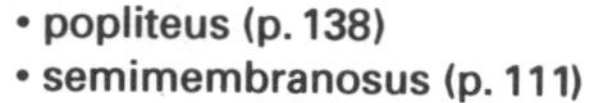

- popliteus (p. 138)
- semimembranosus (p. 111)
- semitendinosus (p. 110)

assisted by:

- sartorius (p. 105)
- gracilis (pp. 105–106)

Study tasks

- Highlight the names of the muscles shown.
- Study the details of each muscle with reference to the page number given.

Popliteus ('muscle at the back of the knee')

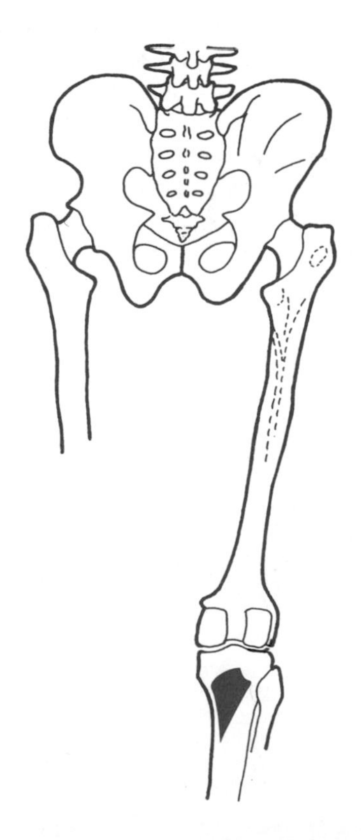

Attachments

- Groove on the lateral femoral condyle.
- Lateral meniscus.
- Posterior surface of the tibia above the soleal line.

Nerve supply

Tibial nerve (L4, L5, S1).

Functions

- Medial rotation of the knee.
- Lateral rotation of the femur (relative to the tibia, rather than as a hip rotator).
- Takes the lateral meniscus posteriorly as the knee flexes.
- Assists flexion of the knee.

Study tasks

- Shade lines between the attachment points and consider the muscle functions.
- Test the muscle on a colleague lying supine with hip and knee flexed to 90°. Support their heel with your hand and resist along the medial border of their foot with your forearm as they try to rotate their leg medially.

Lateral rotation of the knee (0°–40°)

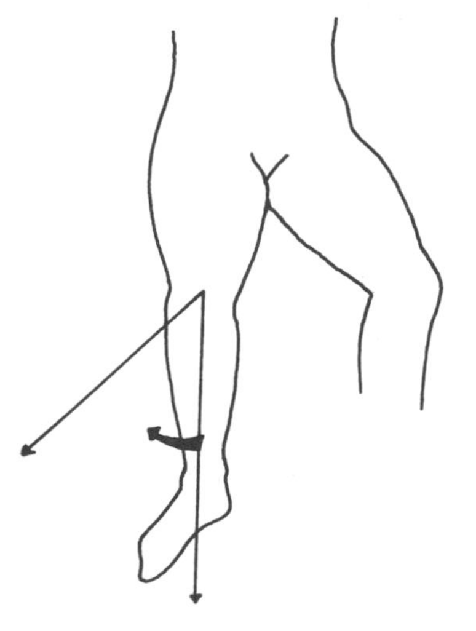

- biceps femoris (p. 109)

Study tasks

- Highlight the name of biceps femoris.
- Study the details of the muscle with reference to the page number given.
- Test the function of biceps femoris as a lateral rotator of the knee on a colleague lying supine with hip and knee flexed to 90°. Support their heel with your hand and resist along the lateral border of their foot with your forearm as they try to rotate their leg laterally.

Chapter 10

The ankle

Introduction

The ankle, or talocrural joint, is the articulation between the tibia and the talus (the uppermost tarsal bone) and the distal end of the fibula. The talus rests on the calcaneus (heel), and the function of the joint, which is a hinge, is to allow the body to spring forwards and upwards when walking and running. The bones of the foot articulate in a rather more complex manner, but the ankle joint is similar to the 'mortise and tenon' arrangement seen in carpentry, with the talus as the tenon. The distal end of the fibula forms the outer ankle bone (lateral malleolus), but is not weight-bearing.

The bones and joints

The bones and bone features of the ankle

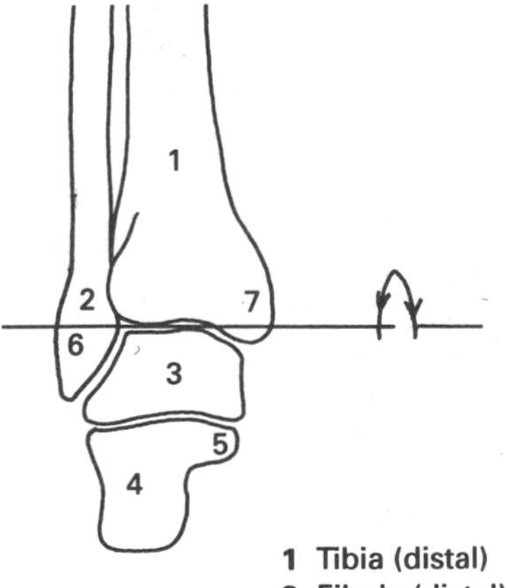

Transverse axis allowing flexion and extension movements only, known as plantar flexion and dorsiflexion (p. 143)

1 Tibia (distal)
2 Fibula (distal)
3 Talus
4 Calcaneus
5 Sustentaculum tali ('ledge for the talus')
6 Lateral malleolus
7 Medial malleolus

Study tasks

- Shade the bones in separate colours.
- Highlight the names of the bone features.

Ligaments and joints of the ankle

Anterior view

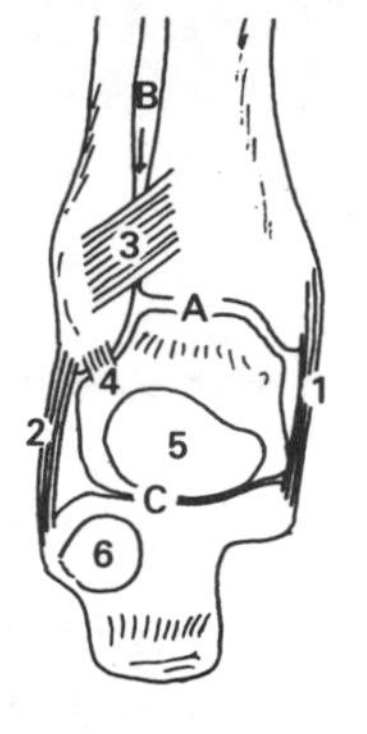

1 Deltoid (medial) ligament
2 Calcaneofibular ligament (part of lateral ligament)
3 Anterior tibiofibular ligament
4 Anterior talofibular ligament
5 Head of talus
6 Facet for cuboid on calcaneus

A Tibiotalar joint (hinge)

B Inferior tibiofibular joint (syndesmosis)

* **C** Subtalar joint (anterior, or talocalcaneonavicular part) (multi-axial)

* This joint permits eversion/inversion of the foot (p. 155)

Posterior view

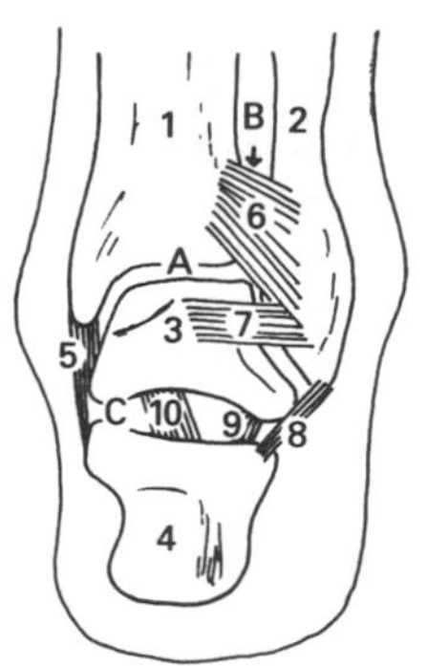

1 Tibia
2 Fibula
3 Talus
4 Calcaneus
5 Deltoid (medial) ligament
6 Posterior tibiofibular ligament
7 Posterior talofibular ligament
8 Calcaneofibular ligament (part of lateral ligament)
9 Cervical ligament (part)
10 Interosseous talocalcanean ligament

A Tibiotalar joint (hinge)

B Inferior tibiofibular joint (syndesmosis)

* **C** Subtalar joint (posterior, or talocalcanean part) (modified multi-axial)

* This joint permits eversion/inversion of the foot (p. 155)

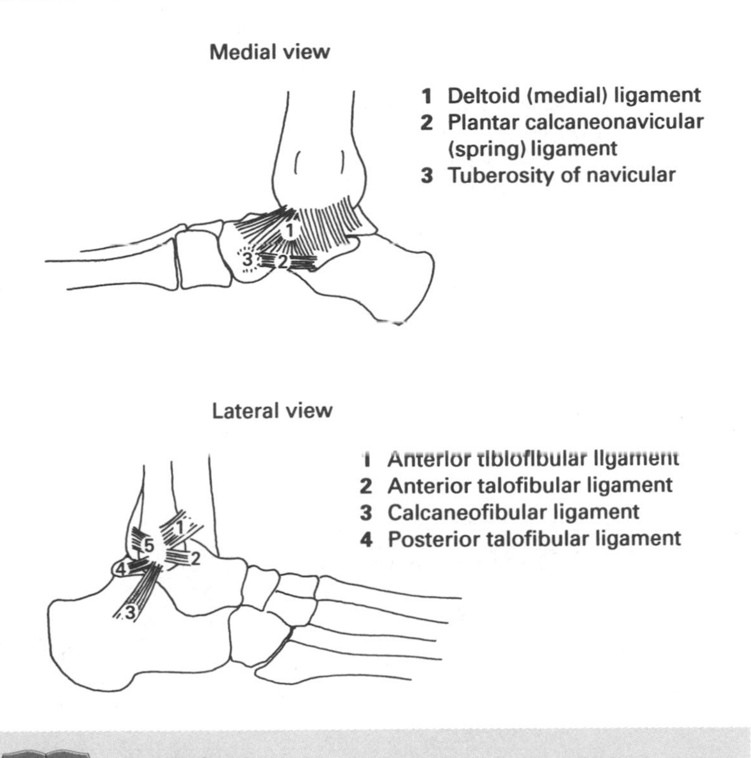

Study tasks

- Highlight the names of the joints and study the details given.
- Shade the ligaments and highlight the features as indicated.
- Palpate the line of the ankle joint by following the tibia (shin) to the point between the ankles where a distinct hollow can be felt. (The tendon of tibialis anterior (p. 144) is prominent if the foot is fully dorsiflexed (see below).)

Movements

Only flexion and extension movements are actively produced by muscles through a transverse axis at the talocrural joint. These movements are called dorsiflexion and plantar flexion.

The foot itself is more mobile, but the talocrural joint does not allow more than a permissive degree of passive abduction/adduction before the collateral ligaments take up the strain, and if excessive, suffer damage. Foot movements are discussed in Chapter 11.

Dorsiflexion

Dorsiflexion is the 'close-pack' position or 'best-fit' between the joint surfaces. It is the optimum position for passive articulation and manipulation of the foot if unwanted movement at the ankle is to be avoided. (See, for example, Hartman (1990).)

Dorsiflexion of the ankle (0°–15°)

- tibialis anterior (p. 144)

assisted by:
- extensor digitorum longus (p. 145)
- extensor hallucis longus (p. 146)
- peroneus tertius (p. 146)

Study tasks

- Highlight the names of the muscles.
- Study the details of each muscle with reference to the page number given.

Tibialis anterior ('front muscle of the shin bone')

Attachments

- ■ Superior/lateral tibia and interosseous membrane.
- ■ Medial surface of medial cuneiform and base of first metatarsal.

Nerve supply

Deep peroneal nerve (L4, L5).

Functions

- ◆ Dorsiflexion of the ankle joint.
- ◆ Inversion of the foot (p. 155).

Study tasks

- Shade lines between the attachments points and consider the muscle functions.
- Rest your heel on the ground, and gently rock your foot from plantar to dorsiflexion while palpating the belly of the muscle. Also identify the muscle tendon as it crosses the ankle joint.
- Test the muscle on a supine colleague whilst holding their foot slightly in plantar flexion and inversion. Restrain the foot while they attempt to dorsiflex.

Extensor digitorum longus ('long extensor muscle of the toes')

Attachments

- Antero-superior fibula, lateral tibial condyle and interosseous membrane.
- Middle and distal phalanges of lateral four toes.

Nerve supply

Deep peroneal nerve (L5, S1).

Functions

- Extension of the interphalangeal and metatarsophalangeal joints of the foot.
- Assists dorsiflexion of the ankle joint.

Study tasks

- Shade lines between the attachment points and consider the muscle functions.
- Palpate the tendon as with tibialis anterior by rocking the heel from plantar to dorsiflexion. Extensor digitorum longus tendon can be felt as a hard edge lateral to tibialis anterior tendon as it crosses the ankle joint line. Flex and extend the toes to palpate its particular action.
- Test the muscle on a supine colleague by restraining their four lateral toes as they try to extend them.

Extensor hallucis longus ('long extensor muscle of the big toe')

Attachments

- Middle part of the anterior fibula and interosseous membrane.
- Base of the distal phalanx of the big toe.

Nerve supply

Deep peroneal nerve (L5, S1).

Functions

- Extension of the big toe.
- Assists dorsiflexion of the ankle joint.

Study tasks

- Shade lines between the attachment points and consider the muscle functions.
- Palpate the tendon as it crosses the ankle joint just lateral to tibialis anterior and medial to extensor digitorum longus, by extending the big toe.
- Test the muscle on a colleague by resisting extension of their big toe.

Peroneus tertius ('the third muscle of the fibula') ('Perone' is Classical Greek for fibula.)

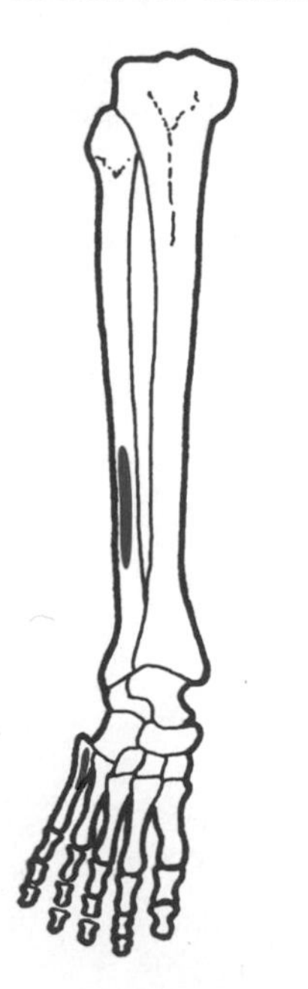

Attachments

- Lower part of the anterior fibula.
- Base of the fifth metatarsal.

Nerve supply

Deep peroneal nerve (L5, S1).

Functions

- Assists eversion of the foot (p. 155).
- Assists dorsiflexion of the ankle joint.

Study tasks

- Shade lines between the attachment points and consider the muscle functions.

Plantar flexion

Plantar flexion of the ankle (0°–55°)

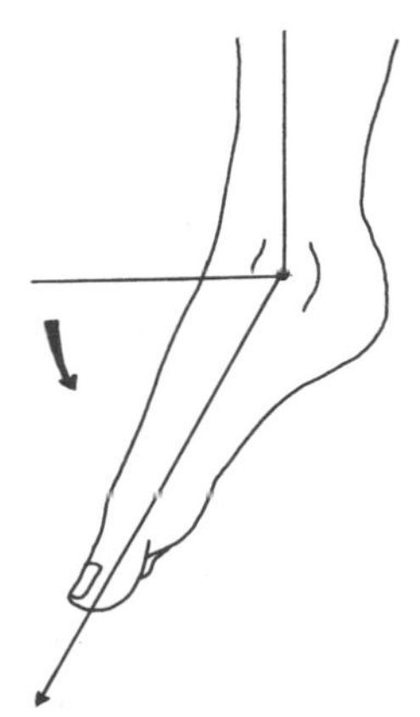

- gastrocnemius (p. 147)
- soleus (p. 148)

assisted by:

- plantaris (p. 149)
- tibialis posterior (p. 149)
- flexor hallucis longus (p. 150)
- flexor digitorum longus (p. 151)

Study tasks

- Highlight the names of the muscles.
- Study the details of each muscle with reference to the page number given.

Gastrocnemius ('belly-shaped muscle of the leg')

Attachments

- Lateral head: on the lateral femoral condyle and lower supracondylar line.
- Medial head: on the popliteal surface of the femur, above the medial condyle.
- Common tendon insertion with soleus and plantaris muscles into the tendo-calcaneus (Achilles tendon).

Nerve supply

Tibial nerve (S1, S2).

Functions

- Plantar flexion of the ankle joint.
- Assists flexion of the knee.

Study tasks

- Shade lines between the attachment points, and consider the muscle functions.
- Test the muscle on a supine colleague by cupping the heel of their foot in your hand and resisting their plantar flexion with your forearm. This is normally a powerful movement, which can also be tested on a colleague by asking them to stand 'on their toes'. It is a useful neurological test of the integrity of S1 motor function.
- Look for the two heads of the muscle on the calf of a colleague with good muscle definition.

Soleus ('muscle which flexes the sole')

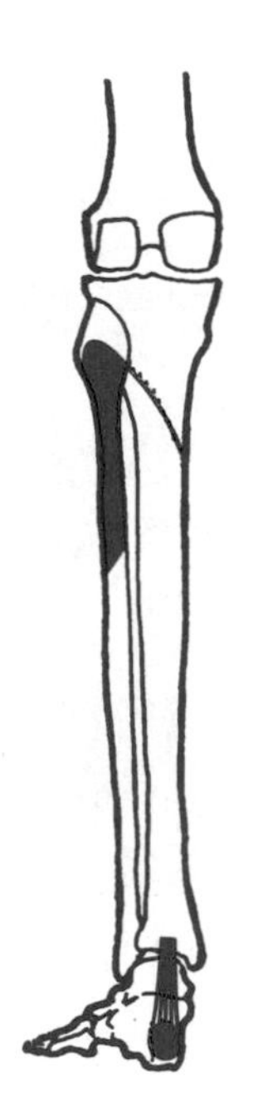

Attachments

- Superior part of the posterior fibula.
- Soleal line on the tibia and deep fascia linking the fibula.
- Common tendon insertion into the tendo-calcaneus (Achilles tendon).

Nerve supply

Tibial nerve (S1, S2).

Functions

- Plantar flexion of the ankle joint.
- Postural role in standing.

Study tasks

- Shade lines between the muscle attachment points and consider the muscle functions.
- Test this muscle which lies deep to gastrocnemius by flexing a colleague's knee to 90° and resist plantar flexion as for gastrocnemius.
- Consider why it is better to test soleus on a flexed knee and gastrocnemius on an extended knee.

Plantaris ('muscle of plantar flexion') (**Note**: this muscle is not always present.)

Attachments

- Popliteal surface of the femur.
- Tendon passes between gastrocnemius and soleus to insert into the medial aspect of tendo-calcaneus.

Nerve supply

Tibial nerve (S1, S2).

Function

- Assists gastrocnemius in plantar flexion.

Study task

- Shade lines between the attachment points and consider the muscle function.

Tibialis posterior ('posterior muscle of the shin bone')

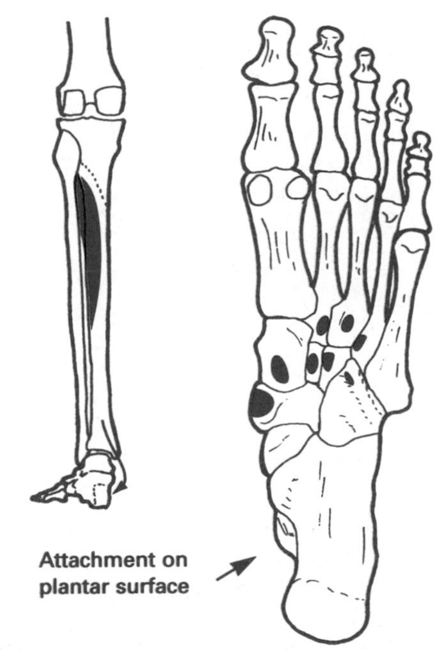

Attachments

- Upper part of the posterior tibia, below the soleal line.
- Upper part of the posterior fibula, and interosseous membrane.
- Wide-ranging insertion into the navicular tuberosity, medial and intermediate cuneiforms (tarsal bones of the foot), and the bases of second, third and fourth metatarsals; occasionally to lateral cuneiform and cuboid.

Nerve supply

Tibial nerve (L4, L5).

Functions

- ♦ Plantar flexion of the ankle joint.
- ♦ Inversion of the foot (p. 155).

Study tasks

- Shade lines between the attachment points and consider the muscle functions.
- Test the muscle on a supine colleague. Slightly invert their foot, cup their heel in your hand and resist plantar flexion with your forearm.
- Palpate the muscle tendon in the hollow between the navicular tubercle and the medial malleolus.

Flexor hallucis longus ('long flexor muscle of the big toe')

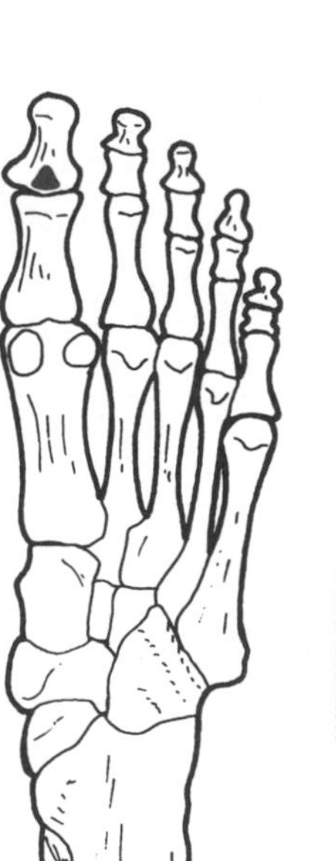

Attachment on plantar surface using sustentaculum tali as a 'pulley'

Attachments

- ■ Lower part of the posterior fibula.
- ■ Base of the distal phalanx of the big toe.

Nerve supply

Tibial nerve (S2, S3).

Functions

- ♦ Flexion of the big toe.
- ♦ Assists plantar flexion of the ankle joint.

Study tasks

- Shade lines between the attachment points and consider the muscle functions.
- Test the muscle on a colleague by resisting flexion of the distal phalanx of their big toe.

Flexor digitorum longus ('long flexor muscle of the toes')

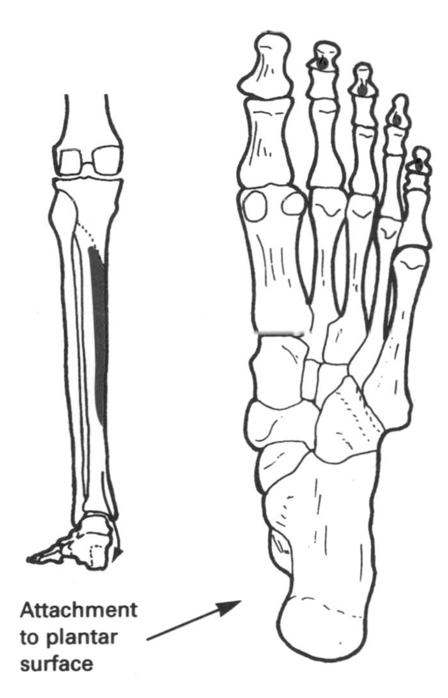

Attachments

- Upper part of the posterior tibia below the soleal line.
- Each digital tendon inserts into the base of a distal phalanx (2–5) after passing through the bifurcated tendons of flexor digitorum brevis (p. 158).

Nerve supply

Tibial nerve (S2, S3).

Functions

- Flexion of the toes.
- Assists plantar flexion of the ankle joint.

Study tasks

- Shade lines between the attachment points and consider the muscle functions.
- Test the muscle on a colleague by resisting flexion of the distal phalanges of the toes.

Chapter 11

The foot

Introduction

The role of the foot is to support the weight of the body in either stationary mode or locomotion. It is a strong but complex arrangement of bones, joints, ligaments and muscles, yet can be the precise and delicate instrument of the dancer as well as the support of the 'yomping' foot soldier.

The toes, like the fingers, are basically synovial hinge joints capable of flexion and extension at the phalanges, but also capable of abduction and adduction through the ellipsoid surfaces of the metatarsophalangeal joints.

However, the real versatility of the foot comes from the articulation between the talus, calcaneus and navicular bones which allow the foot to be turned inwards and outwards in the movements of inversion and eversion, also termed supination and pronation, respectively. When these movements are combined with the plantar and dorsiflexion movements seen at the ankle joint, as well as the accommodatory flexion and extension of the toes, a considerable range of movement is possible. Supination and pronation have less scope in the foot than the hand, because the pivot arrangement seen in the forearm is absent.

The bones and joints

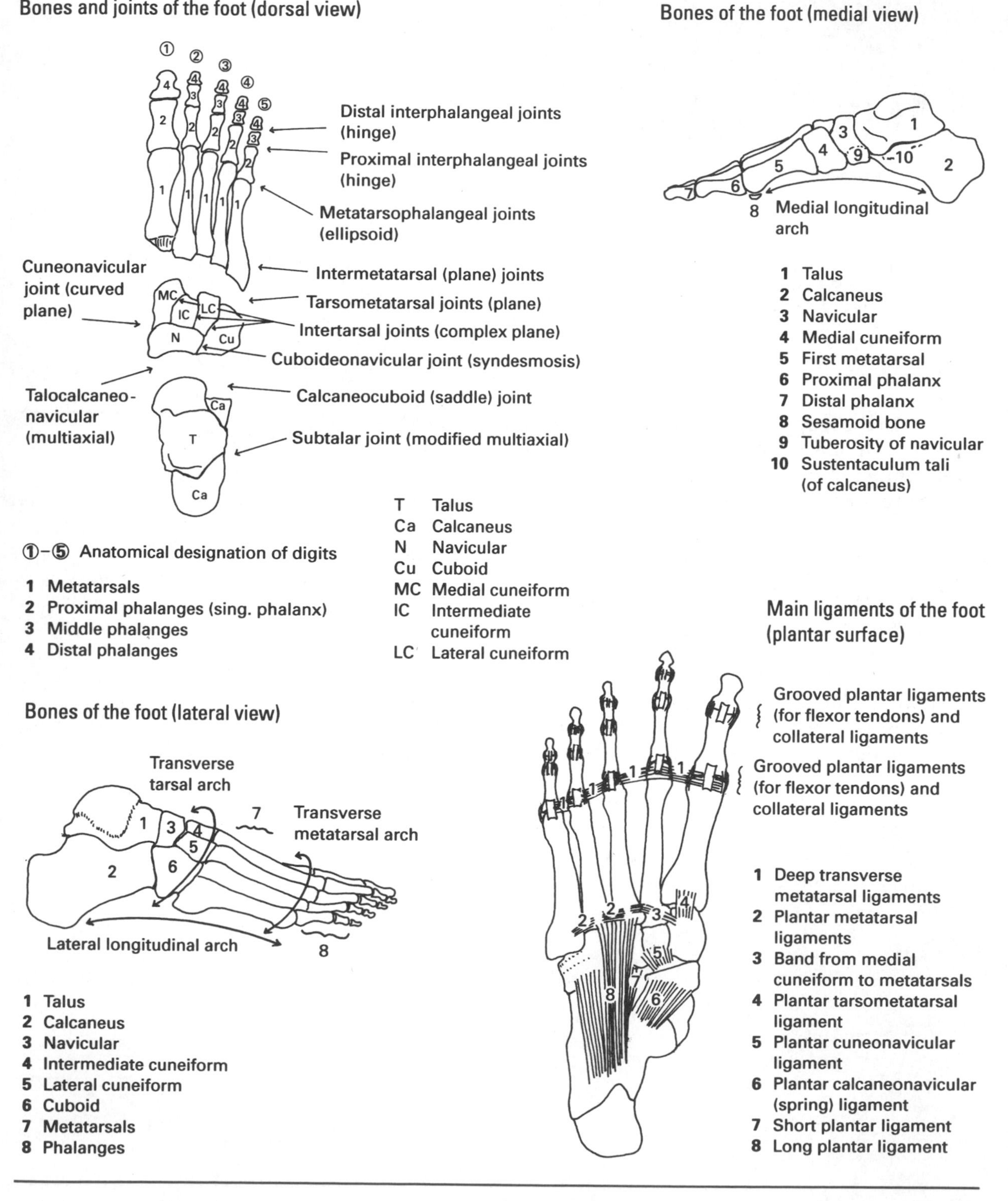

Study tasks

- Shade the bones in separate colours.
- Highlight the names of the bones, joints and ligaments, and study the details given.
- Shade the ligaments and highlight the features as indicated.

Movements of the foot

The movements of inversion and eversion can be produced actively with the foot off the ground. The muscles producing the movements (and the other foot and ankle movements) continue to work when the foot is in contact with the ground, adjusting to uneven surfaces.

Inversion (supination)

Study tasks

- Highlight the names of the muscles.
- Study the details of each muscle with reference to the page number given.

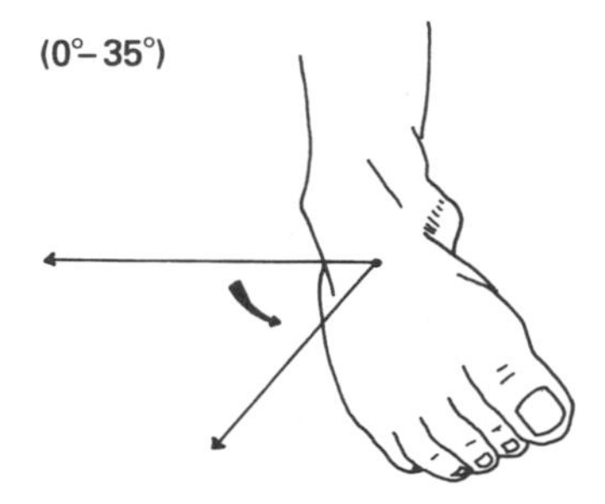

- tibialis anterior (p. 144)
- tibialis posterior (p. 149)

Eversion (pronation)

Study tasks

- Highlight the names of the muscles.
- Study the details of each muscle with reference to the page number given.

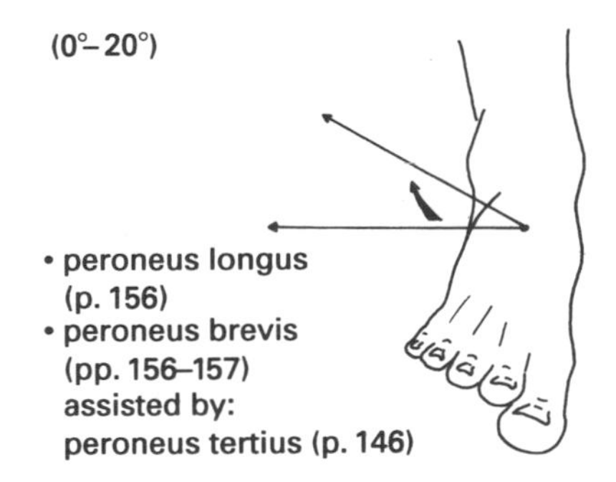

- peroneus longus (p. 156)
- peroneus brevis (pp. 156–157) assisted by: peroneus tertius (p. 146)

Peroneus longus ('long muscle of the fibula')

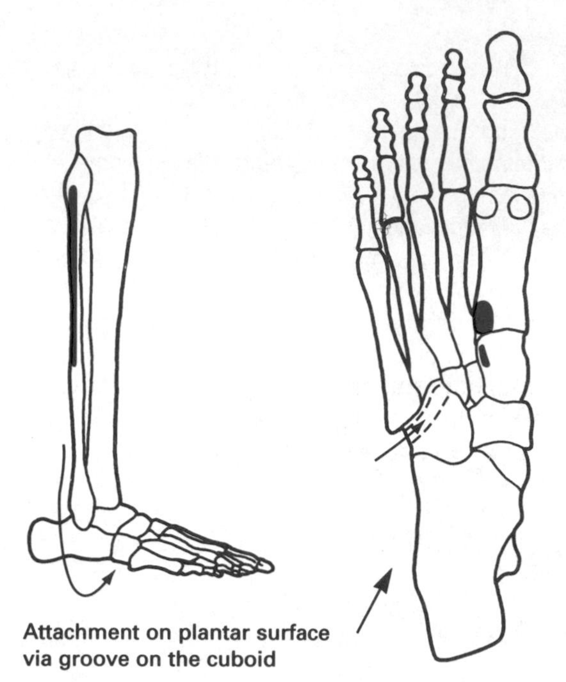

Attachments

- Upper part of the lateral fibula.
- Base of the fifth metatarsal and medial cuneiform via a groove for the tendon on the inferior surface of the cuboid.

Nerve supply

Superficial peroneal nerve (L5, S1, S2).

Functions

- Eversion of the foot.
- The linear pull of the muscle may support the longitudinal and transverse arch of the foot.

Study task

- Shade lines between the attachment points and consider the muscle functions.

Peroneus brevis ('short muscle of the fibula')

Attachments

- Lower part of the lateral fibula.
- Base of fifth metatarsal inferior to peroneus tertius.

Nerve supply

Superficial peroneal nerve (L5, S1, S2).

Function

- ♦ Eversion of the foot.

Study tasks

- Shade lines between the attachment points and consider the muscle functions.
- Test the eversion action of all the peroneal muscles on a supine colleague by placing their foot deliberately in a position of eversion and asking them to hold that position while you attempt to invert the same foot with a gentle and steady pressure.

Movements at the metatarsophalangeal and interphalangeal joints of the toes

The arrangement of these joints is similar to that of the fingers, with the obvious exception that in terms of function the 'big toe' is different from the thumb. Opposition of the 'big toe' is not possible. Extension is greater than flexion, and the foot is adapted to weight-bearing rather than displaying the grasping skills of the hand.

Some of the features of the muscles producing flexion/extension and abduction/adduction will be familiar from Chapter 7 on the hand. However, in the hand, flexion exceeds extension at the metacarpophalangeal joints, whereas in the foot, extension is greater than flexion at the metatarsophalangeal joints.

Flexion

(0°– 40°)

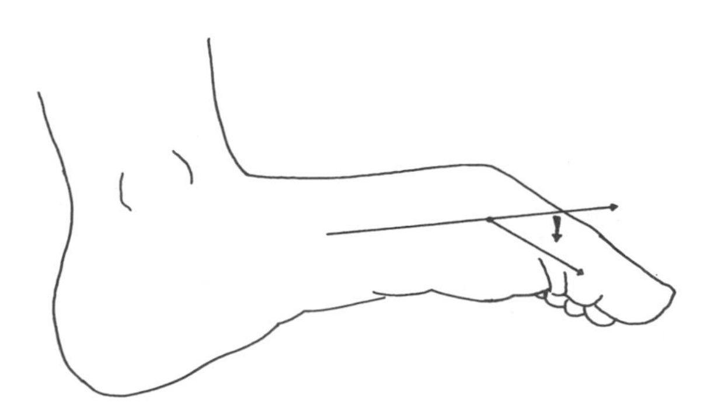

- flexor digitorum longus (p. 151)
- flexor digitorum brevis (p. 158)
- flexor digitorum accessorius (p. 160)
- flexor hallucis longus (p. 150)
- flexor hallucis brevis (p. 161)
- flexor digiti minimi brevis (p. 162)
- lumbricals (p. 158)
- interossei (p. 159)

Study tasks

- Highlight the names of the muscles.
- Study the details of each muscle with reference to the page numbers given.

Flexor digitorum brevis ('short flexor muscle of the toes')

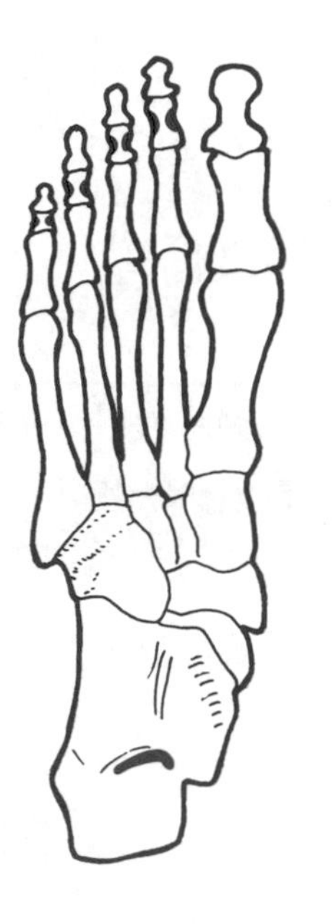

Attachments

- Medial part of tubercle on the calcaneus.
- Plantar surface of the base of second–fifth middle phalanges. The tendons split to allow flexor digitorum longus to reach the distal phalanges (cf. flexor digitorum superficialis, p. 84).

Nerve supply

Medial plantar nerve (S2, S3).

Function

- Flexion of second–fifth toes.

Study tasks

- Shade lines between the attachment points and consider the muscle function.
- Test the muscle on yourself by attempting to pick up a towel or a pencil with your toes.

The lumbricals ('the worm-like muscles')

The line of pull and *modus operandi* is similar to the lumbricals in the hand. The same flexion-with-extension capabilities are theoretically possible in the lumbricals and interossei, but extension at the interphalangeal joints is not really feasible. Since the fine precision movements seen in the hand are not generally required in the foot, the main function of these muscles is effectively to prevent the toes from buckling under when walking.

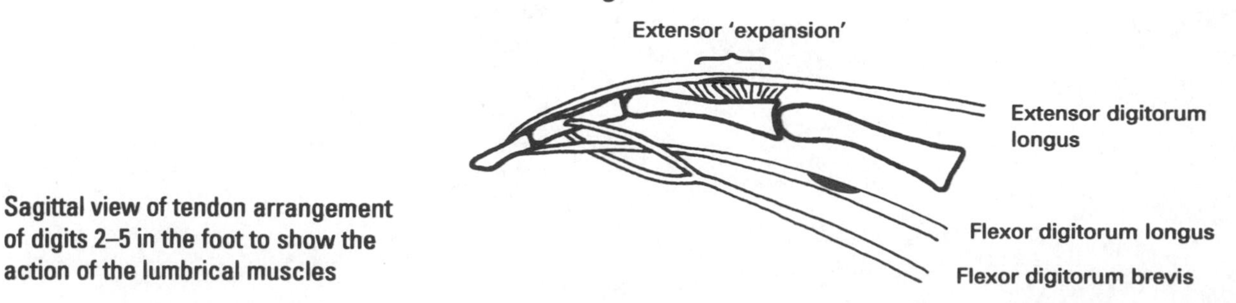

Sagittal view of tendon arrangement of digits 2–5 in the foot to show the action of the lumbrical muscles

Attachments

- Tendons of flexor digitorum longus.
- Medial side of the dorsal expansion of extensor digitorum longus, and thence to the bases of the proximal phalanges.

Nerve supply

Medial and lateral plantar nerves (S2, S3).

Function

- Flexion of second–fifth metatarsophalangeal joints.

Study task

- Shade lines between the attachment points and consider the muscle functions.

The interossei ('the muscles between the bones')

As in the hand, these small muscles occupy the gaps between the 'meta-' bones, but there are only three plantar interossei in the foot because the big toe has its own adductor muscle.

(i) The plantar interossei ('the muscles between the metatarsal bones on the sole')

Attachments

- Medial aspect of third–fifth metatarsals.
- Medial aspect of the base of the proximal phalanx and extensor expansion of the relevant toe.

Nerve supply

Lateral plantar nerve (S2, S3).

Functions

- Flexion of the relevant metatarsophalangeal joints.
- Relative adduction of the third–fifth toes towards the second toe (p. 165).

Study task

- Shade lines between the attachment points and consider the muscle functions.

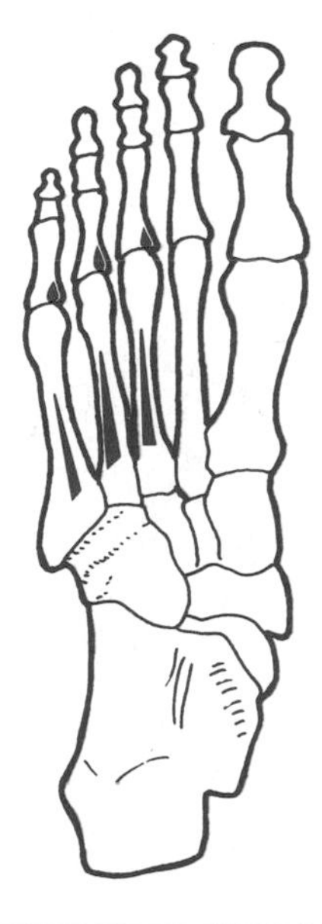

(ii) The dorsal interossei ('the muscles between the metatarsal bones on the upper surface of the foot')

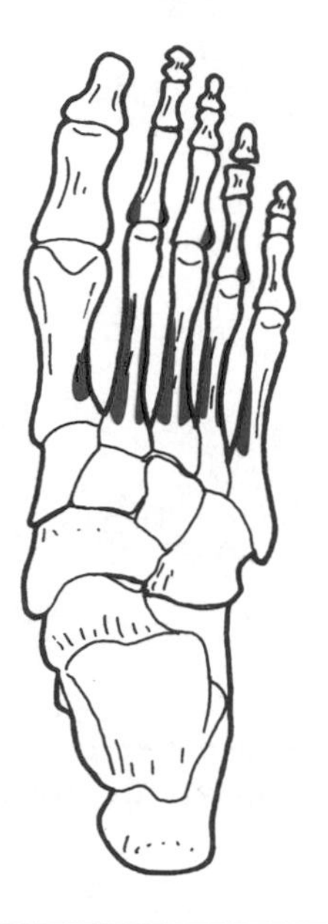

Attachments

- Double attachment points from each side of the metatarsal shafts, except on digits one and five.
- Base of the proximal phalanx and extensor expansion of each relevant toe.

Nerve supply

Lateral plantar nerve (S2, S3).

Functions

- Flexion of the metatarsophalangeal joints.
- Abduction of third and fourth toes (p. 163).

Study task

- Shade lines between the attachment points and consider the muscle functions.

Flexor digitorum accessorius (quadratus plantae) ('accessory flexor muscle of the toes'; 'square-shaped muscle of the sole')

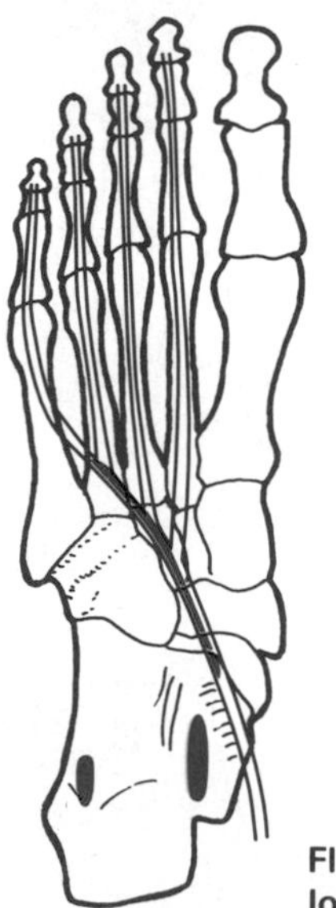

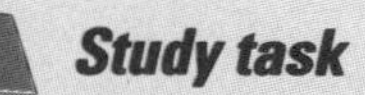

Attachments

- Two heads from medial and lateral calcaneus.
- Lateral margin of flexor digitorum longus tendon.

Nerve supply

Lateral plantar nerve (S2, S3).

Function

- It pulls the obliquely directed tendon of flexor digitorum longus into line so that it acts on the toes more directly.

Study task

- Shade lines between the attachment points and consider the muscle functions.

Flexor hallucis brevis ('short flexor muscle of the big toe')

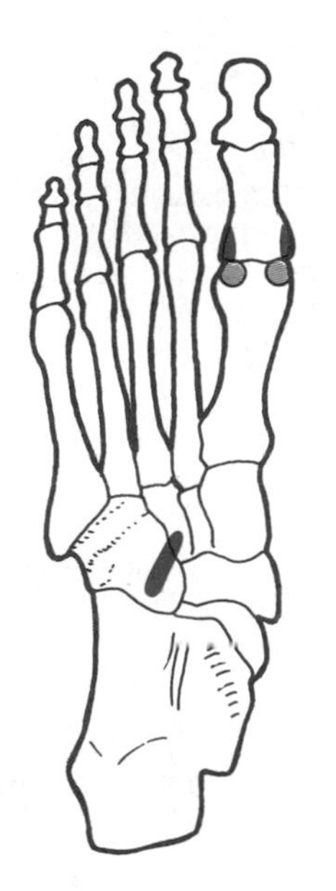

Attachments

- Cuboid.
- Medial side of the base of the proximal phalanx of the big toe.
- Lateral side of the base of the proximal phalanx of the big toe.

(There is a sesamoid bone in each tendon and the tendon of flexor hallucis longus passes between them.)

Nerve supply

Medial plantar nerve (S2, S3).

Function

- Flexion of the big toe at the metatarsophalangeal joint.

Study tasks

- Shade lines between the attachment points and consider the muscle function.
- The muscle may be self-tested by pulling the big toe gently into a degree of extension at the proximal phalanx, and then resisting flexion. It is important that restraint is applied to the proximal phalanx specifically.

Flexor digiti minimi brevis ('short flexor muscle of the little toe')

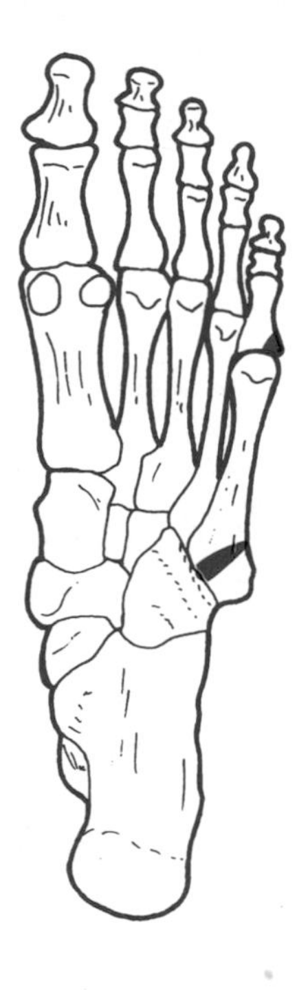

Attachments

- Base of the fifth metatarsal.
- Lateral side of base of proximal phalanx of little toe.

Nerve supply

Lateral plantar nerve (S2, S3).

Function

- Flexes the metatarsophalangeal joint of the little toe.

Study tasks

- Shade lines between the attachment points and consider the muscle function.
- Test the muscle by placing a finger under the proximal phalanx of the little toe and restraining flexion.

Extension

The normal range of extension is greater than flexion at the metatarsophalangeal joints, especially at the first joint.

(0°– 60°)

- extensor digitorum longus (p. 145)
- extensor digitorum brevis (p. 163)
- extensor hallucis longus (p. 146)

Study tasks

- Highlight the names of the muscles.
- Study the details of each muscle with reference to the page number given.

Extensor digitorum brevis ('short extensor muscle of the toes')

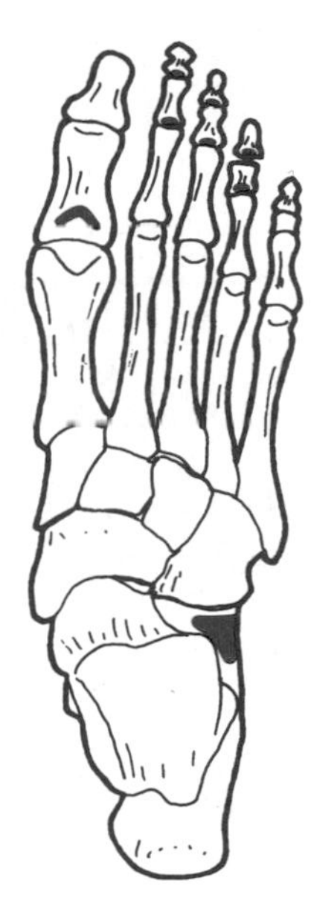

Attachments

- Anterior aspect of upper surface of the calcaneus.
- Base of proximal phalanx of the big toe (this attachment may be called extensor hallucis brevis).
- Extensor expansion of the second–fourth toes (p. 158) and with extensor digitorum longus.

Nerve supply

Deep peroneal nerve (S1, S2).

Function

Extension of the first–fourth toes.

Study tasks

- Shade lines between the attachment points and consider the muscle function.
- Test the muscle by placing two fingers over the dorsal surfaces of the proximal phalanges of the first and second toes and restraining extension.

Abduction

Abduction deviates the toes away from the midline represented by the second toe as a reference point.

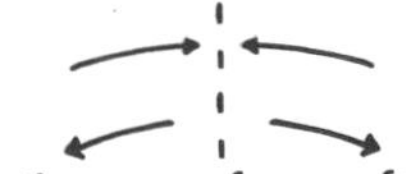

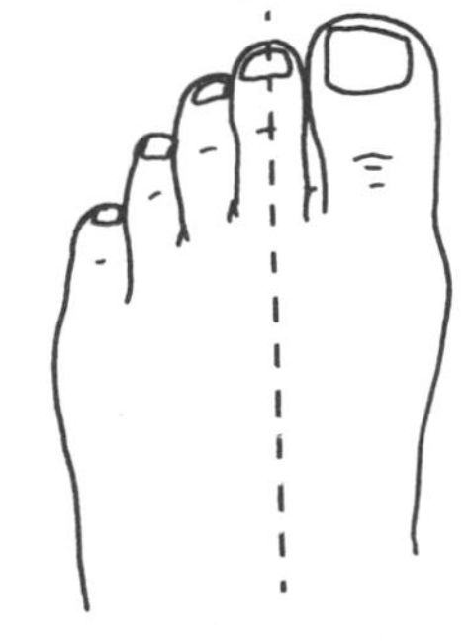

- abductor hallucis (p. 164)
- dorsal interossei (p. 160)
- abductor digiti minimi (p. 164)

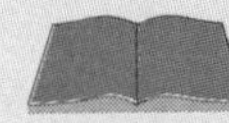

Study tasks

- Highlight the names of the muscles.
- Study the details of each muscle with reference to the page number given.

Abductor hallucis ('abductor muscle of the big toe')

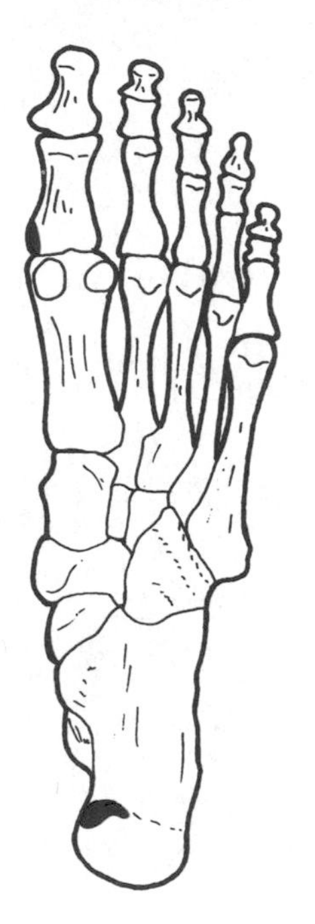

Attachments

- Medial tubercle of the calcaneus and lower edge of the flexor retinaculum.
- Medial side of base of first proximal phalanx.

Nerve supply

Medial plantar nerve (S2, S3).

Functions

- Abduction of the big toe.
- Flexion of the big toe.

Study tasks

- Shade lines between the attachment points and consider the muscle functions.
- Self-test the muscle by placing a restraining finger against the medial side of the big toe and resisting abduction. Slight flexion of the big toe may accompany the movement.

Abductor digiti minimi ('abductor muscle of the little toe')

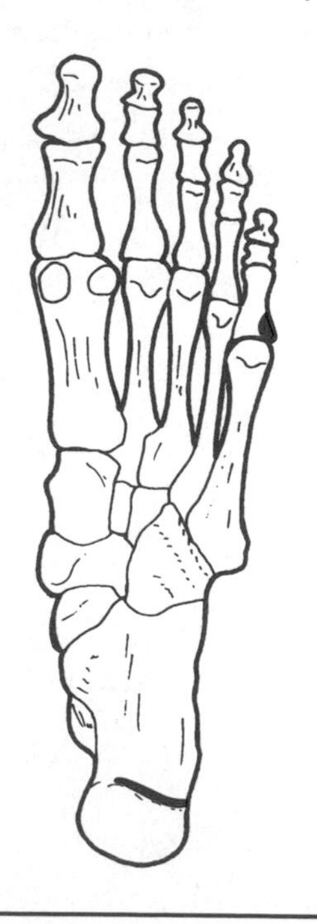

Attachments

- Medial and lateral tubercles of the calcaneus.
- Lateral side of base of fifth proximal phalanx.

Nerve supply

Lateral plantar nerve (S2, S3).

Functions

- Abduction of the little toe.
- Flexion of the little toe.

Study tasks

- Shade lines between the attachment points and consider the muscle functions.
- Test the muscle by placing a finger against the lateral border of the little toe and restrain the action of abduction. Flexion at the metatarsophalangeal joint will also occur.

Adduction

Adduction returns the toes to the midline and is therefore a relative movement from abduction.

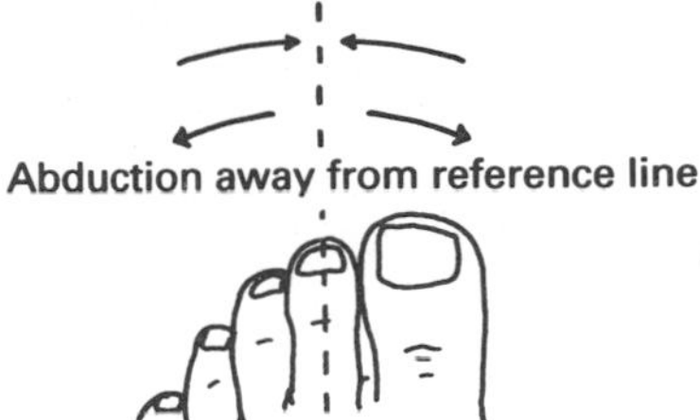

Study tasks

- Highlight the names of the muscles.
- Study the details of each muscle with reference to the page number given.

Adductor hallucis ('adductor muscle of the big toe')

Attachments

- There are two 'heads':
  transverse head: from plantar metatarsophalangeal ligaments of third–fifth toes;
  oblique head: from bases of second–fourth metatarsals and tendon sheath of peroneus longus muscle.
- Common insertion point on lateral base of first proximal phalanx and lateral sesamoid.

Nerve supply

Lateral plantar nerve (S2, S3).

Functions

- Adduction of the big toe.
- The transverse head draws the metatarsal bones closer together and therefore helps to maintain the transverse arch.

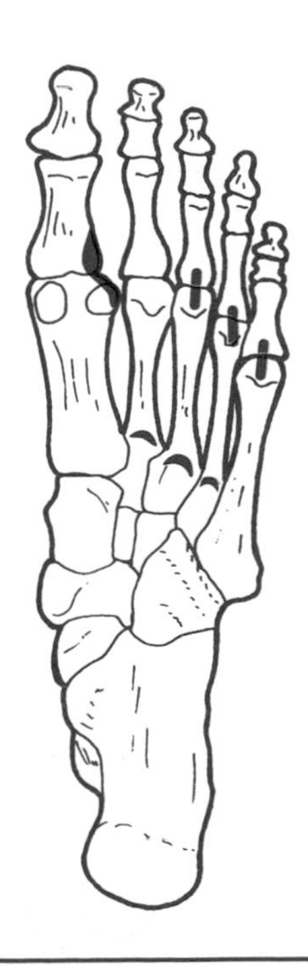

Study tasks

- Shade lines between the attachment points and consider the muscle functions.
- Self-test the muscle by placing one foot on the ground and while trying to press the big toe into adduction try to create a hollow under the transverse arch. This is a useful strengthening exercise for the transverse arch, and may be enhanced by attempting the same exercise with a tennis ball under the transverse arch.

Chapter 12

The respiratory (and abdominal) muscles

Introduction

Respiration involves the movement of a wide area of the body which includes the ribs, sternum and thoracic spine, as well as the diaphragm, abdomen and cervical spine. The respiratory muscles are usually classified as muscles of inspiration (breathing in) and expiration (breathing out). There is a further subdivision of those used at all times, including quiet breathing (primary muscles), and those used when extra effort is necessary (accessory muscles) (Kapandji, 1987). In this context, the abdominal muscles are an important group of synergistic accessory muscles of expiration. They have other important functions which will be considered as well, but can be conveniently placed in the latter part of this chapter.

The mechanism of breathing

The main respiratory movements are focused on the rib cage and thorax.

Inspiration

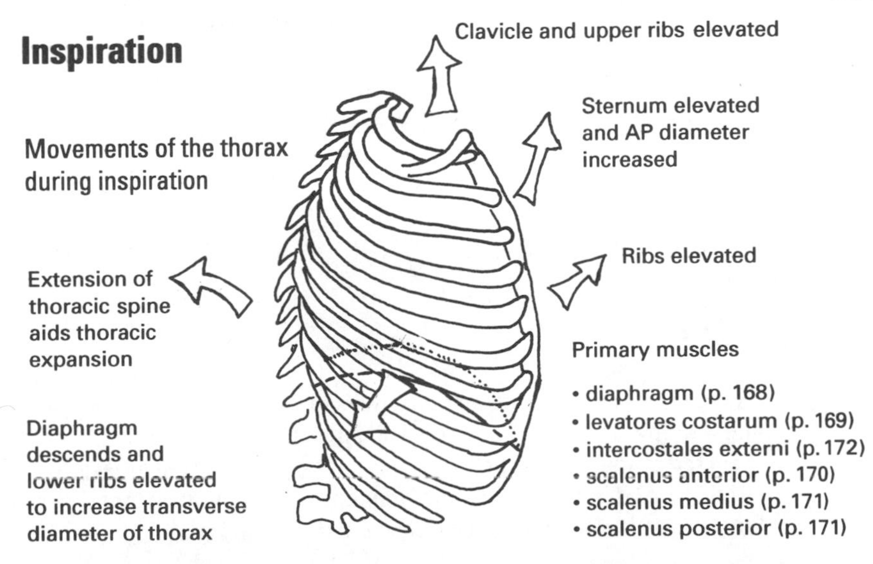

Movements of the thorax during inspiration

Primary muscles

- diaphragm (p. 168)
- levatores costarum (p. 169)
- intercostales externi (p. 172)
- scalenus anterior (p. 170)
- scalenus medius (p. 171)
- scalenus posterior (p. 171)

Accessory muscles

- iliocostalis cervicis (pp. 201–202)
- sternocleidomastoid (p. 173)
- pectoralis major (p. 174)
- pectoralis minor (p. 175)
- latissimus dorsi (p. 176)
- serratus anterior (p. 177)
- serratus posterior superior (p. 178)
- quadratus lumborum (p. 178)

Study tasks

- Highlight the names of the features indicated, including the names of the muscles.
- Study the details of each muscle with reference to the page number given.
- Observe the difference between quiet breathing and deep breathing in a colleague, and yourself. It may be helpful to place one hand on the abdomen and the other on the ribs while doing this.
- Feel the difference between upper and lower rib deep breathing by placing one hand just below the clavicle and the other on the lateral aspect of the lower ribs. The upper ribs move upwards in a manner often described as 'pump handle' action, while the lower ribs flare outwards in a manner described as 'bucket handle' action (Snell, 1995).

The primary muscles of inspiration

These muscles are always used, including during quiet breathing.

The diaphragm ('the partitioning muscle')

This is a muscular dome whose apex separates the thoracic from the abdominal cavity. Its attachments are quite complex, and additional details may be obtained from a standard anatomy textbook.

Attachments

- Xiphisternum (retrosternal surface).
- Lower six ribs and costal cartilages (inner surface).
- Medial and lateral tendinous arches (arcuate ligaments) covering the psoas major and quadratus lumborum muscles.
- Tendinous slips (crura) covering L1–3 and blending with the anterior longitudinal ligament.
- Muscle fibres from the attachments above converge onto the central tendon at the apex of the muscular dome.

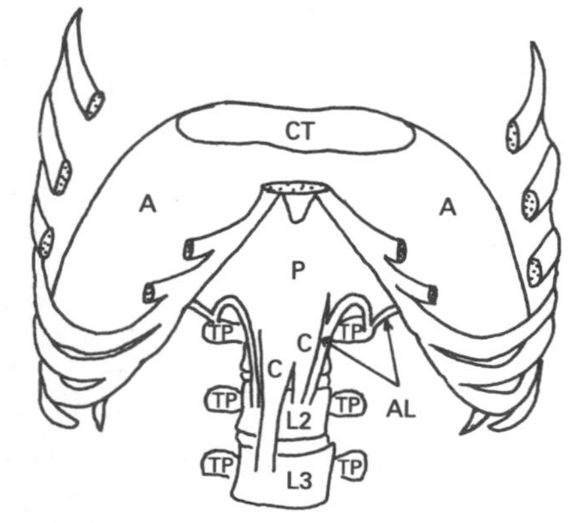

CT Central tendon
A Anterior diaphragm
P Posterior diaphragm
C Crura
AL Arcuate ligaments
TP Transverse process

Nerve supply

Phrenic nerve (C3, C4, C5: motor supply); ventral rami of T6–12 (sensory supply).

Functions

- As the main muscle of inspiration it pulls the central tendon down and thereby increases the vertical dimension of the thorax. Descent is limited by mediastinal stretch superiorly and abdominal pressure inferiorly. At this point the central tendon becomes in effect the 'origin', and further contraction elevates the lower ribs.

- It separates the thoracic and abdominal cavities.

Diaphragm descends on inspiration and is opposed by abdominal contents

Further contraction of peripheral diaphragm elevates lower ribs

- It helps to produce the effect known as the 'Valsalva manoeuvre'. This is produced by inhaling deeply, closing the glottis so that the air cannot escape, and raising intra-abdominal pressure by contracting the abdominal muscles. Weightlifters employ this manoeuvre to provide anterior support for the spine.

Study tasks

- Draw arrows from the central tendon downwards to the attachment points on the ribs anteriorly, and to the lumbar vertebrae posteriorly, to indicate the domed shape of the muscle.
- The action of the muscle may be experienced by simply taking a deep breath and holding it for a few seconds. This should not be attempted by anyone with cardiovascular or respiratory disorders.
- Focus your attention on your own diaphragm by lying supine and breathing quietly with hands resting over lower ribs and abdomen. The abdomen rises gently on inspiration. Lung capacity is greater in the lower thorax and breathing exercises often focus on this. It is frequently referred to as 'abdominal breathing'.

Levatores costarum (sing. levator costae) ('muscles which lift the ribs')

These muscles lie posteriorly, and superficial to the external intercostal muscles.

Attachments

- TPs of C7–T11.
- Outer surface of the rib below between the tubercle and the 'angle' of the rib.

Nerve supply

Dorsal rami of the corresponding thoracic nerves.

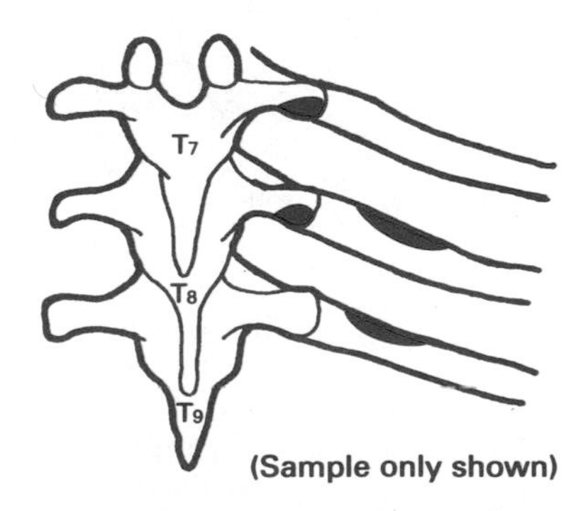

(Sample only shown)

Functions

- The direction of attachment would suggest inspiratory action in raising the ribs posteriorly.
- Unilaterally, slight sidebending and rotation of the thoracic spine to the same side.

Study task

- Shade lines between the attachment points (only a sample section of the thoracic spine is shown) and consider the muscle functions.

Scalenus anterior ('front triangular muscle with unequal sides')

Attachments

- Anterior tubercles of TPs C3–6.
- Scalene tubercle on the first rib.

Nerve supply

Ventral rami of C4–6.

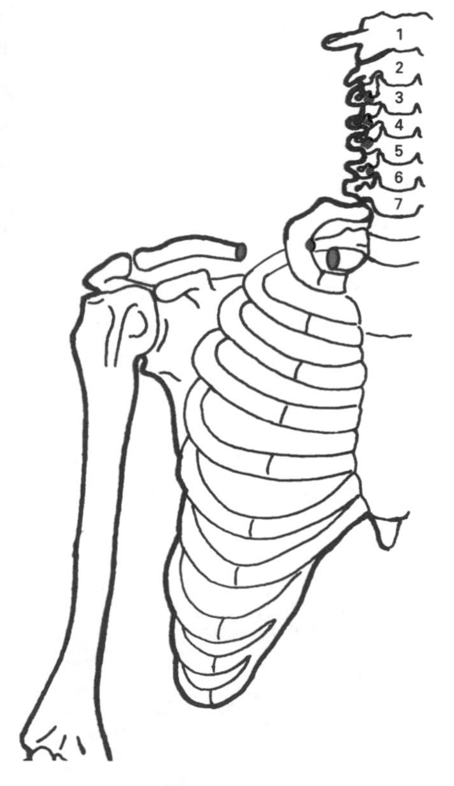

Functions

- Bilaterally, assists elevation of the first rib in inspiration when the cervical spine is fixed.
- Bilaterally, assists flexion of the cervical spine if ribs are fixed.
- Unilaterally, sidebending and slight rotation of the cervical spine to the opposite side if the first rib is fixed. Rotation occurs because the anterior tubercles of the vertebrae lie posterior to the scalene tubercle on the first rib. (See also the action of sternocleidomastoid, p. 173.)

Study tasks

- Shade lines between the attachment points, and consider the muscle functions.
- Palpate the muscles on a seated colleague, by standing behind and resting your hands gently on the clavicle. Palpate the supraclavicular fossa with the flat part of the fingertips as your colleague takes a deep breath. You should be able to feel scalenus anterior (and medius) contract, raising the first rib.

Attachments

- Posterior tubercles of TPs C2–7.
- Upper surface of the first rib behind the subclavian groove. (The subclavian artery lies between scalenus anterior and medius.)

Nerve supply

Branches from ventral rami of C3–8.

Functions

- Bilaterally, assists elevation of the first rib in inspiration when the cervical spine is fixed.
- Bilaterally, assists flexion of the cervical spine if the ribs are fixed.
- Unilaterally, sidebending of the cervical spine if first rib is fixed.

Study tasks

- Shade lines between the attachment points and consider the muscle functions.
- Palpate the muscle action as for scalenus anterior.

Scalenus medius ('middle triangular muscle with unequal sides')

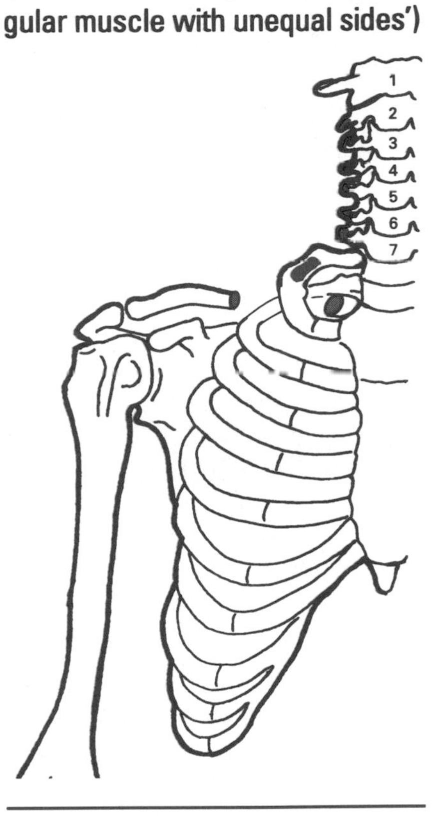

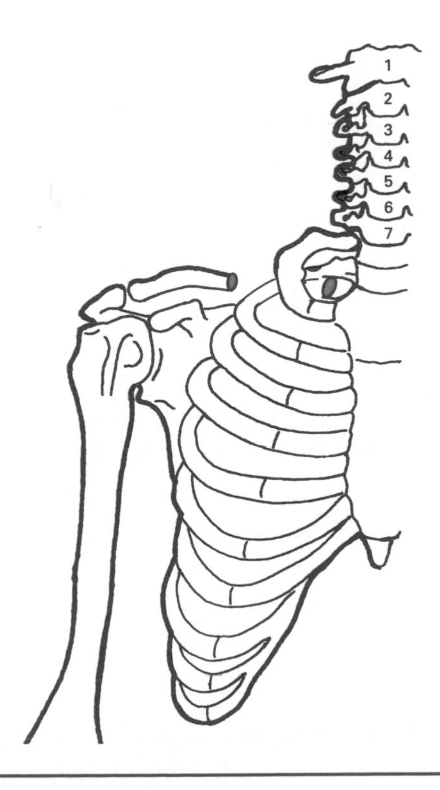

Scalenus posterior ('muscle at the back with unequal sides')

Attachments

- Posterior tubercles of TPs C4–6.
- Upper surface of the second rib.

Nerve supply

Branches from ventral rami of C6–8.

Functions

- Bilaterally assists elevation of the second rib in inspiration when the cervical spine is fixed.
- Bilaterally assists cervical flexion if the ribs are fixed.
- Unilaterally sidebending of the cervical spine if the second rib is fixed.

Study task

- Shade lines between the attachment points and consider the muscle functions.

Intercostales externi ('outer muscles between the ribs')

There are eleven of these muscles on each side, between the ribs.

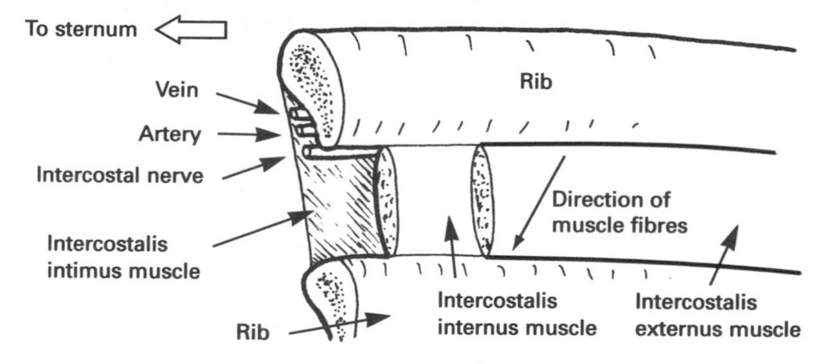

Attachments

- Lower surface of the rib above from the tubercle to the costal cartilage.
- Upper surface of the rib below at an oblique angle inferiorly and laterally at the back, medially at the front.

Nerve supply

The adjacent intercostal nerves.

Function

- Elevation of each rib below in inspiration.

Study task

- Shade lines between the attachment points in the direction indicated.

The accessory muscles of inspiration

These are muscles which are recruited during deep inspiration, and most of them have other functions as well.

Sternocleidomastoid ('muscle attached to sternum, clavicle and mastoid process')

This muscle has two heads which attach separately to the clavicle and sternum. The clavicular fibres lie deeper.

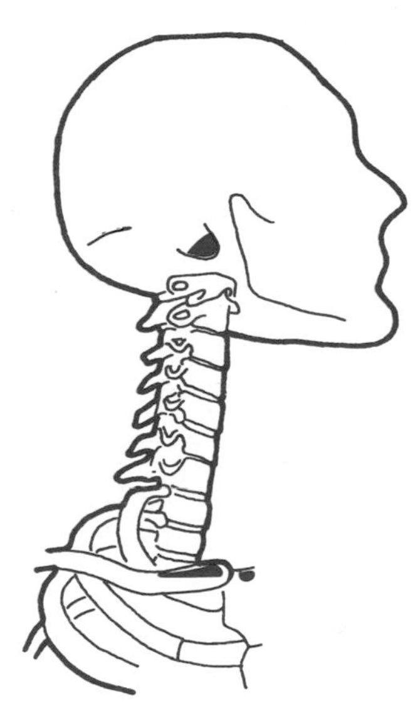

Attachments

- Upper part of the manubrium sterni.
- Medial part of the clavicle.
- Mastoid process (clavicular fibres).
- Lateral superior nuchal line (sternal fibres).

Nerve supply

Cranial nerve (XI) and ventral rami of C2, C3, C4.

Functions

- Bilaterally, elevation of the thorax in deep inspiration if the head is held stationary.
- Bilaterally, flexion of the head and neck if the thorax is fixed.
- Unilaterally, sidebending of the head, with rotation to the opposite side.

Study tasks

- Shade lines between the attachment points and consider the muscle functions. (See *Gray's Anatomy* (1995) for precise details about muscle arrangement.)
- Test the muscle by palpating the sternal attachment of the left side between your left thumb and index finger. Next place the palm of your right hand against the right side of your face and try to turn your head to the right.
- If you resist with your right hand you will feel the left sternal head contract. The effect is enhanced by sidebending the head to the left, and can be clearly observed in a mirror as rotation is increased.
- Vary the degrees of sidebending and rotation in the above test, and try to define the roles of the sternal and clavicular parts in each movement. Would you expect this from the attachment points on the diagram?

Pectoralis major ('larger muscle of the chest')

(**Note:** With reference to respiration, only the sternocostal part is shown here. See pp. 32–33 for further details of this muscle.)

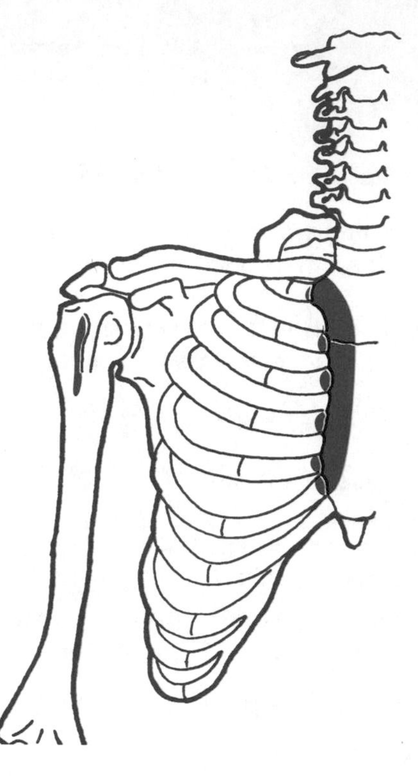

Attachments

- Costal cartilages 1–6 and sternum (sternocostal part).
- Lateral lip of bicipital groove of humerus.

Nerve supply

Lateral and medial pectoral nerves (C7, C8, T1 for the sternostal part only).

Function

- As an accessory muscle of deep inspiration, the upper six ribs are raised if the upper arm is elevated. This is commonly seen after exertion when the hands are clasped behind the head (see also p. 12).

Study tasks

- Shade lines between the attachment points and consider the respiratory function of the muscle.
- Observe the action of the muscle by standing with the shoulders exposed in front of a mirror and raising your arms fully above your head. Clasp your hands so that the shoulders are internally rotated, and take a deep breath. Stretch your arms higher as you do so, and you will see the inferior border of the muscle clearly defined. Notice how it changes position as the arm is raised.

Pectoralis minor ('smaller muscle of the chest')

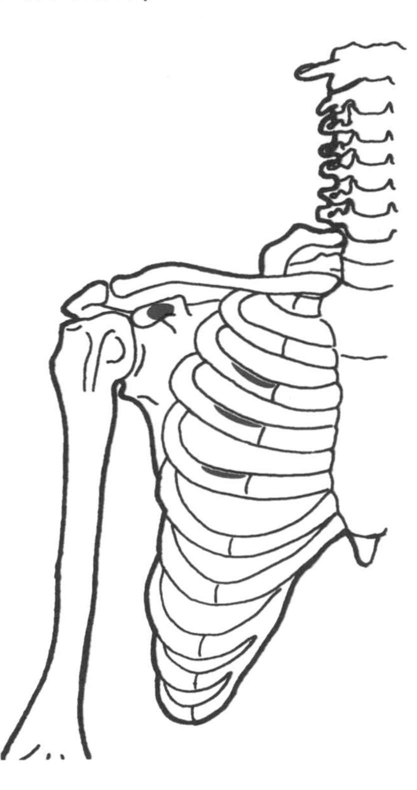

Attachments

- Ribs 3, 4, and 5 near their costal cartilages.
- Coracoid process of the scapula.

Nerve supply

Medial and lateral pectoral nerves (C6, C7, C8).

Functions

- As an accessory muscle of deep inspiration, it assists the sternocostal part of pectoralis major by raising ribs 3–5 on an elevated arm.
- For further functions of this muscle see p. 29.

Study tasks

- Shade lines between the attachment points, and consider the respiratory function of the muscle.
- The muscle lies deep to pectoralis major but may be tested theoretically in the same way as pectoralis major above for its respiratory functions only.

Latissimus dorsi ('widest muscle of the back')

The widest muscle of the back is also attached to the ribs, scapula, and thence to the humerus, like pectoralis major, where it also acts as a medial rotator and adductor muscle.

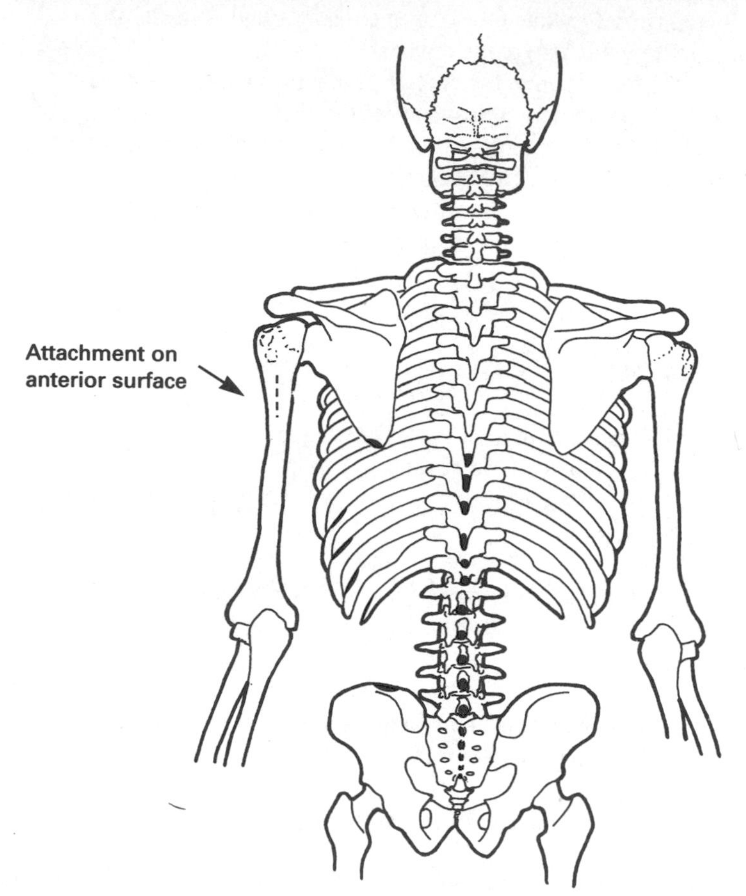

Attachments

- External iliac crest and thoracolumbar fascia.
- SPs of T7–S5.
- Lower four ribs (tendinous slips).
- Inferior angle of the scapula.
- Floor of the bicipital groove (biceps tendon groove of the humerus).

Nerve supply

Thoracodorsal nerve (C6, C7, C8).

Functions

- As an accessory muscle of deep inspiration it raises the lower four ribs when the shoulder girdle is raised on an elevated arm.
- For further functions of this muscle see pp. 37 and 182.

Study task

- Shade lines between the attachment points and consider the respiratory functions of this muscle.

Serratus anterior ('saw-toothed muscle at the front')

Only the lower fibres are effective in elevating the ribs in deep inspiration.

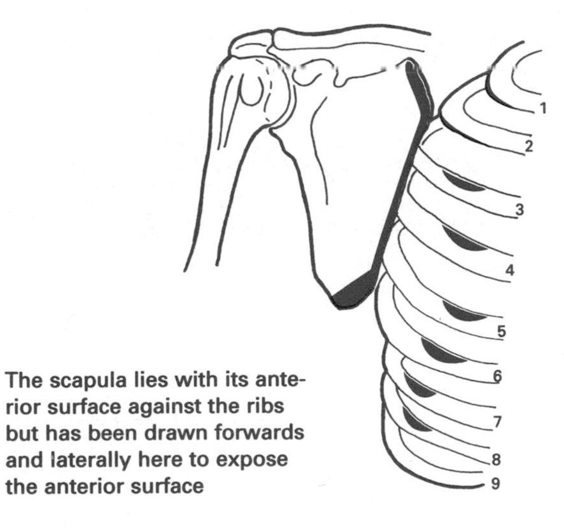

The scapula lies with its anterior surface against the ribs but has been drawn forwards and laterally here to expose the anterior surface

Attachments

- Lower fibres on ribs 6, 7, 8 (the upper fibres attach to ribs 1–5).
- Anterior surface of medial border of the scapula.

Nerve supply

Long thoracic nerve (C5, C6, C7).

Functions

- As an accessory muscle in deep inspiration it raises ribs 6, 7, and 8, if the scapula is fixed on an elevated arm.
- For the other functions of this muscle see p. 28.

Study task

- Shade lines between the attachment points and consider the respiratory function of the muscle.

Serratus posterior superior ('upper saw-toothed muscle at the back')

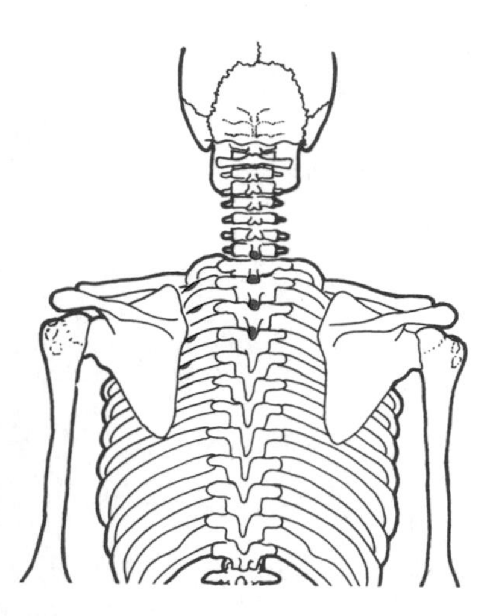

Attachments

- SPs of C7–T3.
- Outer surface of ribs 2–5.

Nerve supply

Intercostal nerves 2–5.

Function

- The attachment points suggest that ribs 2–5 will be elevated in deep inspiration, but there is some uncertainty over this (*Gray's Anatomy*, 1995).

Study task

- Shade lines between the attachment points and consider the muscle function.

Quadratus lumborum ('square-shaped muscle of the low back')

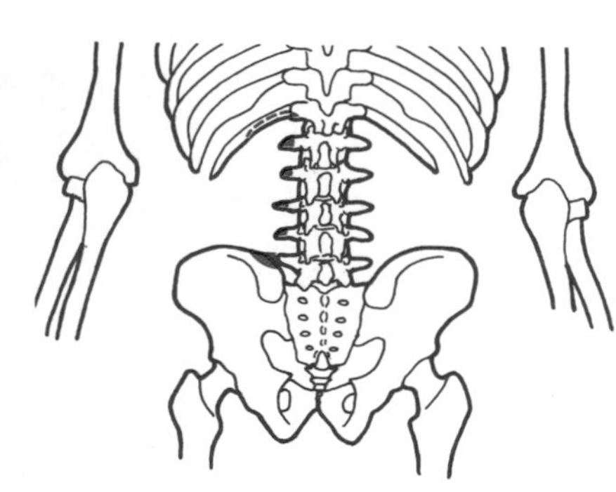

Attachments

- Iliolumbar ligament and posterior aspect of the iliac crest.
- Inferior border of the twelfth rib and TPs of T12 and L1–4.

Nerve supply

Ventral rami of T12; L1–3, (4).

Functions

- As an accessory muscle of inspiration it fixes the twelfth rib, providing a stable border for the descending diaphragm.
- Bilaterally, is assists extension of the lumbar spine.
- Unilaterally, it assists sidebending of the lumbar spine.

Study tasks

- Shade lines between the attachment points and consider the functions of the muscle.
- Test the muscle by placing a colleague in the prone position with their extended legs slightly sidebent. Palpate with one hand below their twelfth rib while your other hand prevents them from straightening their body. This test is limited only to the sidebinding function.

Expiration

In expiration the exhalation of air from the lungs occurs as a result of natural recoil, gravity and the graded relaxation of the muscles of inspiration. As with inspiration, there are a number of muscles whose attachments appear to be helpful in the respiratory process and yet whose role is still not fully substantiated. A number of muscles would seem to assist quiet expiration, while forced expiration may require expulsive effort which recruits the powerful abdominal muscles anteriorly and latissimus dorsi posteriorly.

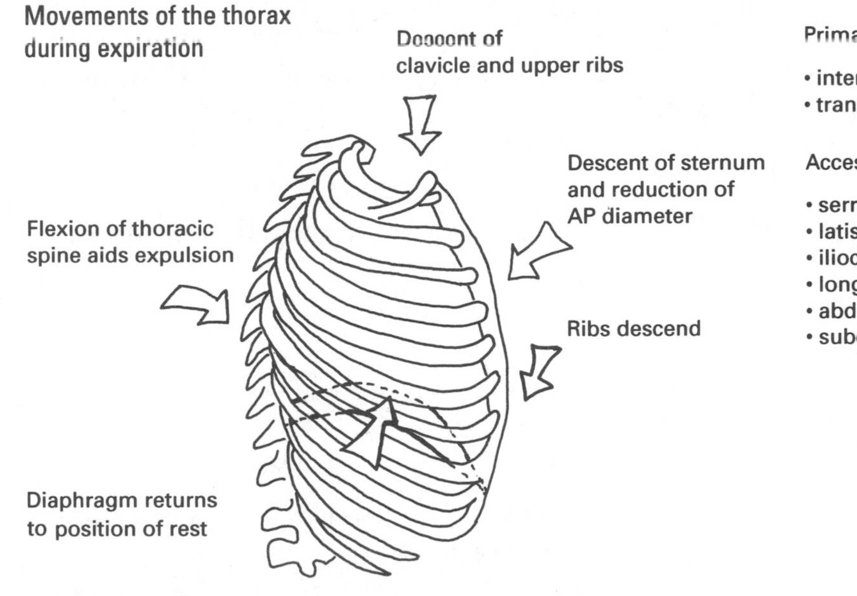

Primary muscles

- intercostales interni (pp. 179–180)
- transversus thoracis (pp. 180–181)

Accessory muscles

- serratus posterior inferior (p. 181)
- latissimus dorsi (p. 182)
- iliocostalis lumborum (p. 203)
- longissimus thoracis (p. 201)
- abdominal muscles (p. 182)
- subcostales (p. 180)

Study tasks

- Highlight the names of the features indicated including the names of the muscles.
- Study the details of each muscle with reference to the page number given.

The primary muscles of expiration

These muscles are always used, including during quiet breathing.

Intercostales interni ('inner muscles between the ribs') (Eleven on each side.)

Note: Deep to this muscle lies the **intercostalis intimus** muscle ('deepest muscle between the ribs') which runs in the same direction and appears to perform the same function. It occupies the middle section of the intercostal rib space, but is poorly developed in the upper ribs. The **subcostales** ('the muscles under the ribs') are also deep and variable fasciculi which are thought to help depress the ribs in expiration. There is, however, uncertainty about the precise role of some of these muscles (*Gray's Anatomy* 1995).

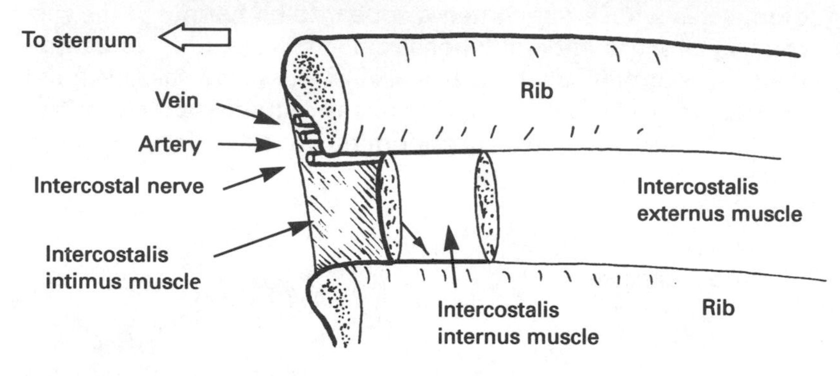

Attachments

- Inferior surface of each rib deep to the external intercostal muscles.
- Superior border of the rib below.

Nerve supply

The corresponding intercostal nerves.

Function

- As a muscle of expiration by drawing the ribs closer together after inspiration has taken place.

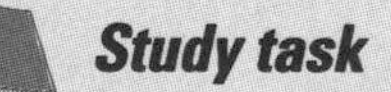

Study task

- Shade lines obliquely between the attachment points, perpendicular to the direction of external intercostal muscles (p. 172) in the direction indicated.

Transversus thoracis (sternocostalis) ('muscle crossing the thorax/attached to ribs and sternum')

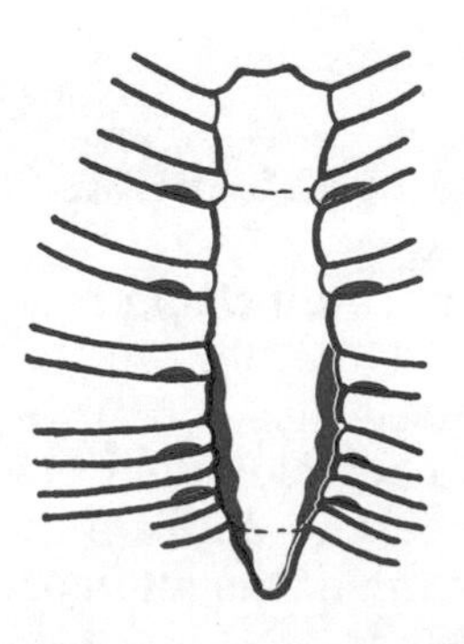

This muscle lies on the inner surface of the anterior thorax.

Attachments

- Lower inner surface of the sternum and xiphisternum.
- Variably to the inner surfaces of costal cartilages 2–6.

Nerve supply

The corresponding intercostal nerves.

Function

- Draws the costal cartilages down during expiration.

Study task

- Shade lines between the attachment points and consider the muscle function.

The accessory muscles of expiration

These muscles are recruited during deep breathing or when forceful effort is required which may be quite sudden and expulsive, as in coughing and sneezing.

Serratus posterior inferior ('lower saw-toothed muscle at the back')

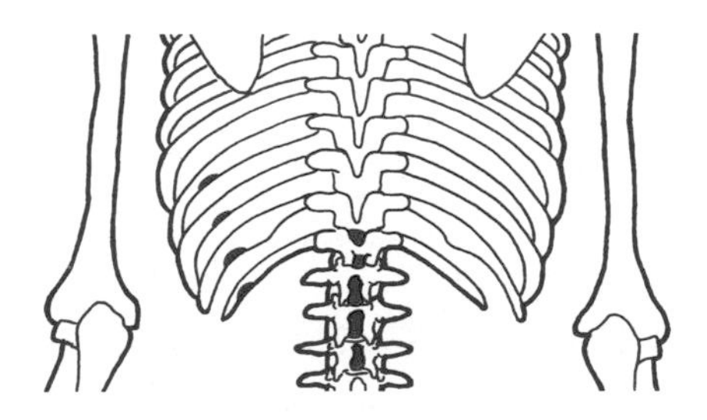

Attachments

- SPs of T11–L3.
- Outer surfaces of ribs 9–12.

Nerve supply

Ventral rami of T9–12.

Functions

- Uncertain, but its attachments suggest that it depresses ribs 9–12 in expiration.
- It may also serve to tighten the thoracolumbar fascia (Dowdney, 1993).

Study task

- Draw lines between the attachment points and consider the muscle functions.

Iliocostalis lumborum and longissimus thoracis

These are erector spinae muscles which, because of their rib attachments, are able to depress the ribs and may therefore be recruited as accessory muscles of expiration.

Study task

- See pp. 201 and 203 for details of these muscles and consider their respiratory function.

Latissimus dorsi

This muscle has already been described as an accessory muscle of deep inspiration as it is able to draw the ribs and scapula upwards on a raised arm (p. 176). However, it is also possible to envisage the lower fibres drawing the ribs downwards, and this appears to happen in sudden expulsive movements, such as coughing.

Study tasks

- Revise the details of this muscle on p. 176 and consider its respiratory functions.
- Palpate the forced expiratory functions of this muscle on a colleague by asking them to cough and palpate the thoracolumbar attachments as they do so.

The abdominal muscles

The superficial muscles of the abdomen

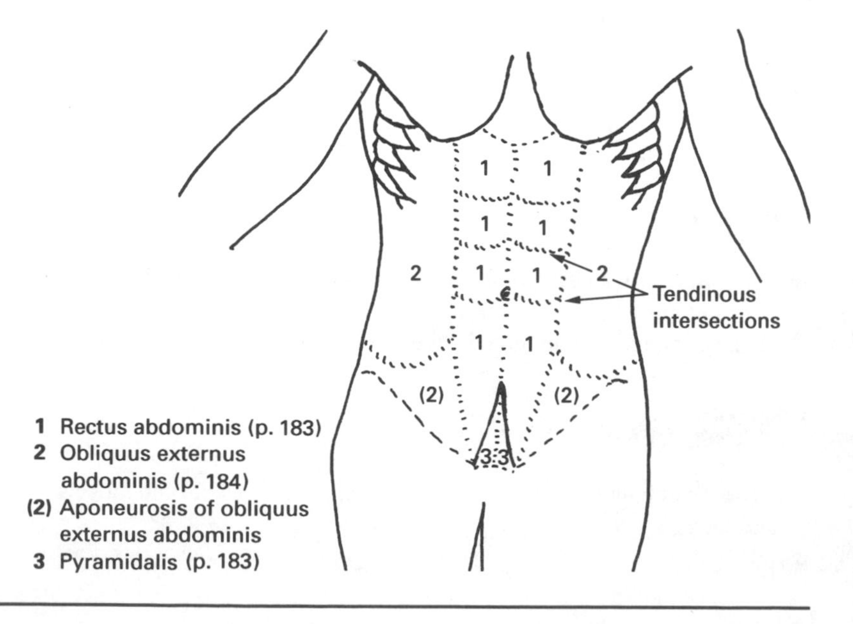

Simplified transverse section of 'rectus sheath' at umbilical level

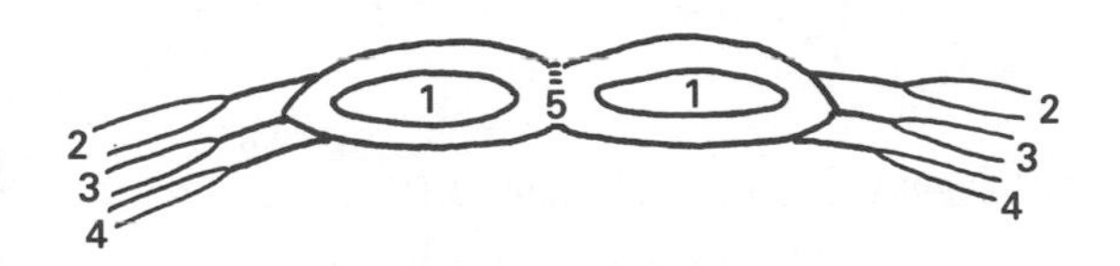

1 Rectus abdominis (p. 183)
2 Obliquus externus abdominis (p. 184)
3 Obliquus internus abdominis (p. 184)
4 Transversus abdominis (p. 185)
5 Linea alba

Study tasks

- Shade the superficial abdominal muscles in separate colours.
- Shade the cross-section of the abdominal muscles in separate colours.
- Highlight the names of the muscles and features shown.
- Study the details of each muscle with reference to the page number given.

Rectus abdominis ('straight supporting muscle of the abdomen')

This consists of two long strap muscles separated by a raphe known as the linea alba ('white line') and surrounded by the 'rectus sheath' and aponeuroses of the oblique abdominal muscles.

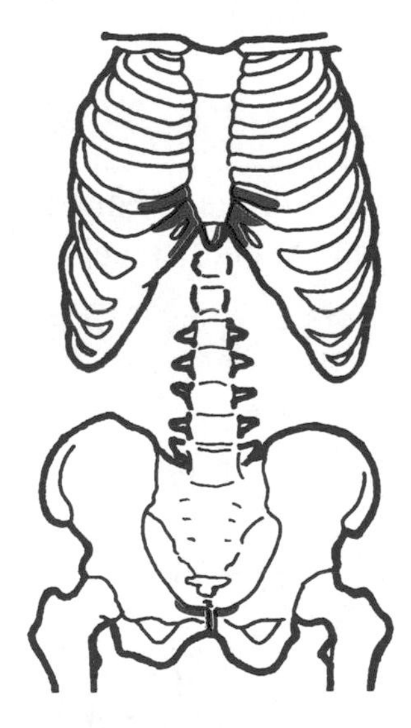

Attachments

- Crest of the pubis and symphysis pubis.
- Costal cartilages of ribs 5–7 and lateral xiphisternum.

Nerve supply

Ventral rami of T5–12.

Study task

- Shade lines between the attachment points of the muscle as shown.

Pyramidalis ('small pyramid-shaped muscle')

A small triangular muscle, sometimes absent, which lies anterior to rectus abdominis. Its function appears to be to tighten the linea alba during abdominal movement. See diagram on p. 182.

Attachments

- Anterior pubis and symphysis pubis.
- Linea alba, midway between pubis and umbilicus.

Nerve supply

Ventral ramus of T12 (subcostal nerve).

Obliquus externus abdominis ('externally angled muscle')

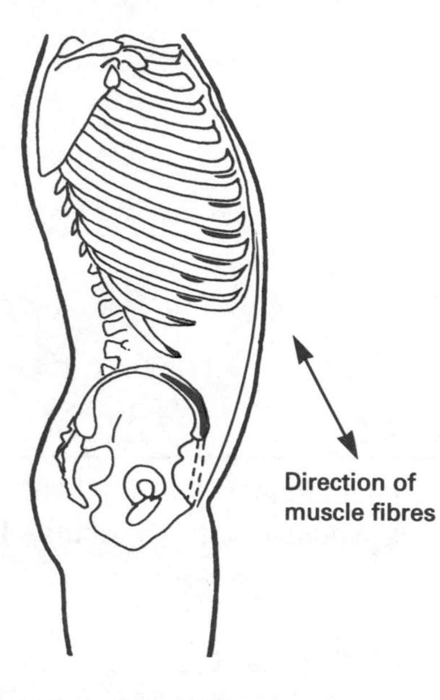

Attachments

- Inferior surface of ribs 5–8 anteriorly and ribs 9–12 laterally.
- Anterior part of the iliac crest and linea alba via a broad aponeurosis.

Nerve supply

Ventral rami of T7–12.

Study task

- Shade lines between the attachment points of the muscle as shown.

Obliquus internus abdominis ('internally angled muscle')

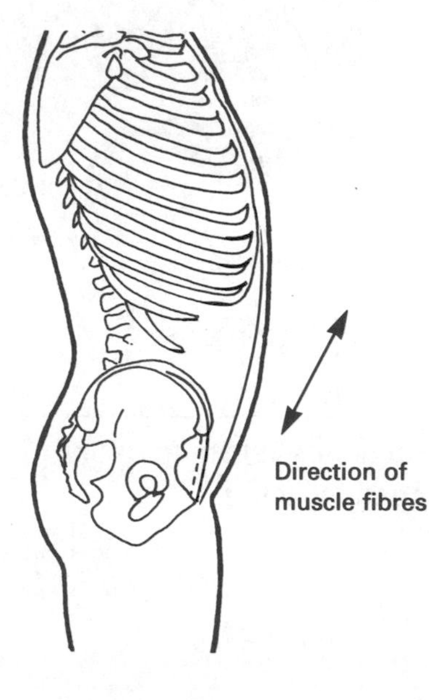

This muscle is thinner and lies deep to obliquus externus.

Attachments

- Lateral part of the inguinal ligament, lumbar fascia and iliac crest.
- Crest of pubis and linea alba via an aponeurosis.
- Inferior surface of ribs 7–12.

Nerve supply

Ventral rami of T7–L1.

Study task

- Shade lines between the attachment points of the muscle as shown.

Transversus abdominis ('muscle with fibres crossing the abdomen')

This is the innermost muscle deep to obliquus internus.

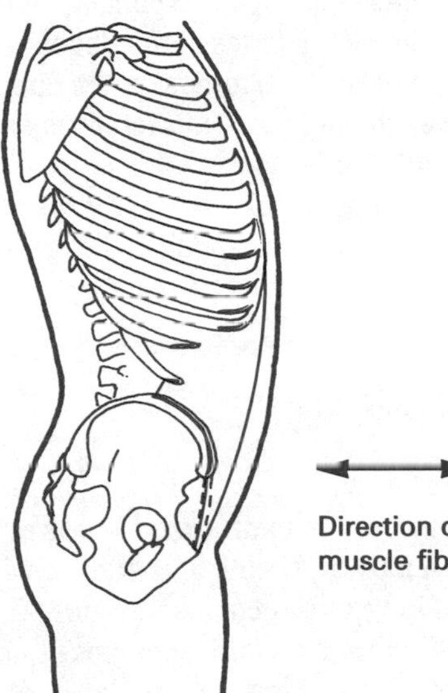

Attachments

- Lateral part of inguinal ligament; thoracolumbar fascia; anterior iliac crest; costal cartilages 7–12.
- Linea alba and the aponeurosis passing posterior to rectus abdominis.

Nerve supply

Ventral rami of T7–L1.

Study task

- Shade lines between the attachment points as shown.

Functions of the abdominal muscles

- As accessory muscles of expiration all the abdominal muscles act together by raising intra-abdominal pressure and expelling air from the lungs. This may be done either gradually, or in an explosive manner such as coughing and sneezing.
- As anterior muscles supporting the abdominal viscera, they also provide resistance to the descending diaphragm. This allows the diaphragm to contract further, elevating the lower ribs.
- As anterior muscles of support, they also provide reinforcement for the vertebral column during lifting. The procedure adopted may require a combination of deep inspiration followed by closure of the glottis (i.e. holding one's breath) and effective support from the muscles of the pelvic diaphragm (p. 121). This is known as the 'Valsalva manoeuvre'.

- Flexion of the lumbar spine from the supine position.
- Rotation of the lumbar spine especially from the supine position due to the direction of the fibres of the obliquus internus on the same side, and the obliquus externus on the opposite side. Rotation to the right involves the right obliquus internus and the left obliquus externus, and vice versa.
- The attachment points of a strong rectus abdominis will oppose anterior tilt of the pelvis, 'abdominal sag' and a tendency to hyperextension of the lumbar spine.

Study tasks

- Test rectus abdominis on a supine colleague whose arms are crossed so that their hands are on opposite shoulders: their elbows then act as a pointer. Their knees and hips should be flexed to reduce the hip flexing action of iliopsoas. They should try to sit up by pointing their elbows towards their knees, as you gently restrain them. The tendinous intersections shown on p. 182 may be visible.
- Sitting up straight tests rectus abdominis and pyramidalis. If the hands are placed behind the neck and the elbows allowed to drop outwards, the obliquus muscles can be tested by taking each elbow alternately to the opposite knee. This may be performed as a self-test, or on a colleague under supervision. If you are uncertain as to which obliquus muscle is being tested, revise the section on these muscles.

Chapter 13

The muscles of the vertebral column

Introduction

The vertebral column is a multi-articular pillar which is segmented and flexible. It has the following functions:

- support
- shock absorption
- protection
- leverage
- metabolic functions (including the manufacture of blood cells from bone marrow)
- physical expression of posture.

The physical expression of posture

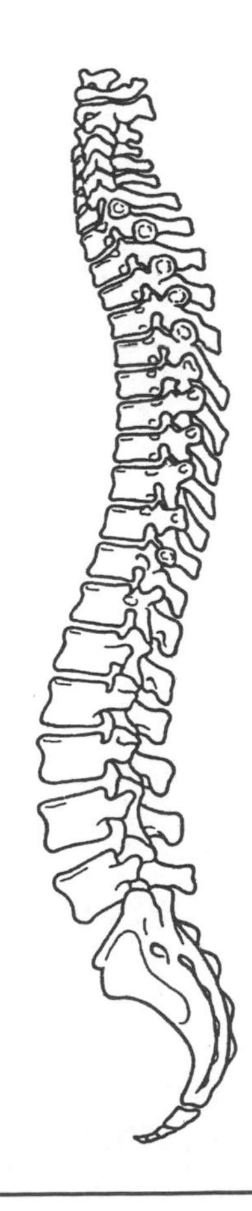

The structure of the individual vertebrae variously reflect these functions. There are seven cervical, twelve thoracic and five lumbar vertebrae, but the basic functional structure of any individual vertebra reflects a common purpose, except for C1 and C2 which are highly specialized.

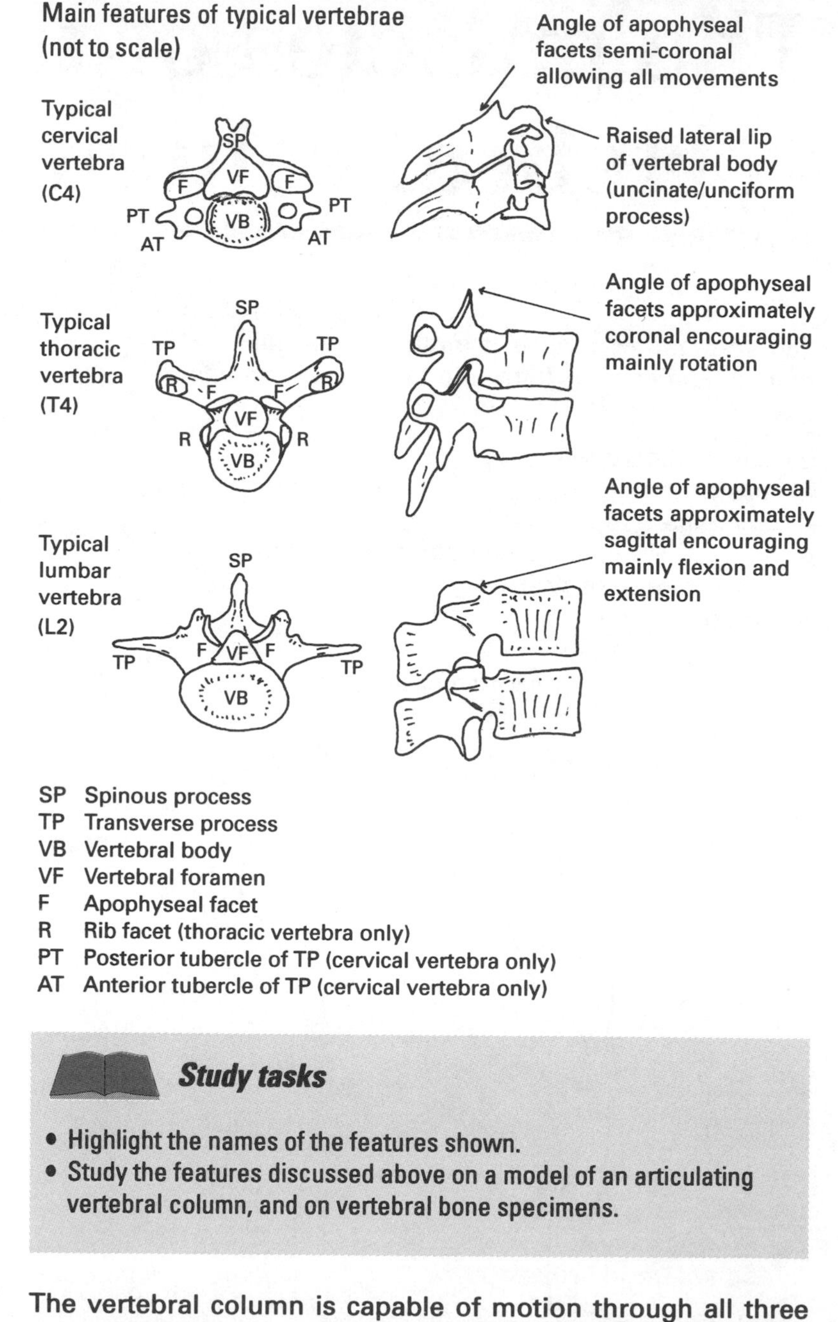

Study tasks

- Highlight the names of the features shown.
- Study the features discussed above on a model of an articulating vertebral column, and on vertebral bone specimens.

Movements of the vertebral column

The vertebral column is capable of motion through all three axes of movement (see pp. 6–7) allowing flexion/extension, sidebending (lateral flexion) and rotation. Circumduction (not shown here) is also possible as a gross movement, due to the summation of the other movements. These movements are represented throughout the vertebral column, but the shape and orientation of the individual apophyseal facet joints which guide the movements provide different emphasis at different levels.

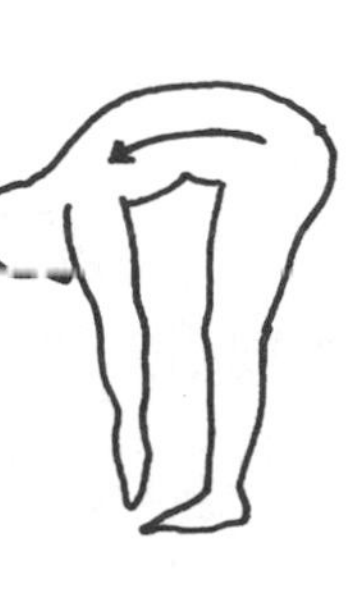

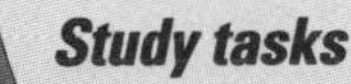

Study tasks

- Highlight the details shown.
- Observe the individual movements of the vertebral column on a healthy colleague who does not suffer from back pain.

The role of the muscles

The vertebral muscles provide support and protection, and leverage and movement for the vertebral column and body generally.

The muscles are attached mainly to the bony levers formed by the transverse and spinous processes of the individual vertebrae. The flexor muscles are attached anteriorly, and the extensor muscles are attached posteriorly.

Sidebending is usually the result of the flexor and extensor muscles acting on one side only. Rotation is produced either to the same side or to the opposite side, depending upon the individual muscle attachment points. Details of the muscles which produce rotation and sidebending (lateral flexion) are given on p. 212.

The anterior vertebral muscles

These muscles predominantly flex the spine when they contract bilaterally. However, it should be noted that unilateral contraction may produce sidebending with rotation, sometimes to the opposite side (stated in the text as appropriate). Gravity and a gradual lengthening of the posterior extensor muscles allow flexion to take place from the standing position, but the anterior muscles actively flex the spine from the supine position.

Flexion of the head and cervical spine

(angle may be measured from
the horizontal 'plane of bite', 0°–75°)

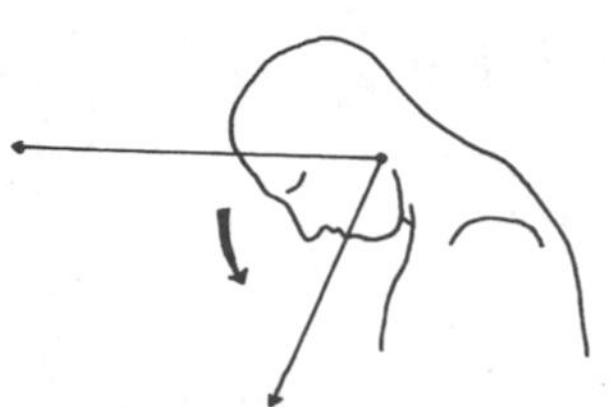

- longus colli (p. 190)
- longus capitis (p. 191)
- sternocleidomastoid (p. 173)
- rectus capitis anterior (p. 192)
- scalenes (pp. 170–172)

Study tasks

- Highlight the names of the muscles.
- Study the details of each muscle with reference to the page number given.

Longus colli ('long muscle of the neck')

This muscle consists of three parts, and the attachments of each part will be considered separately. (Only one side is shown.)

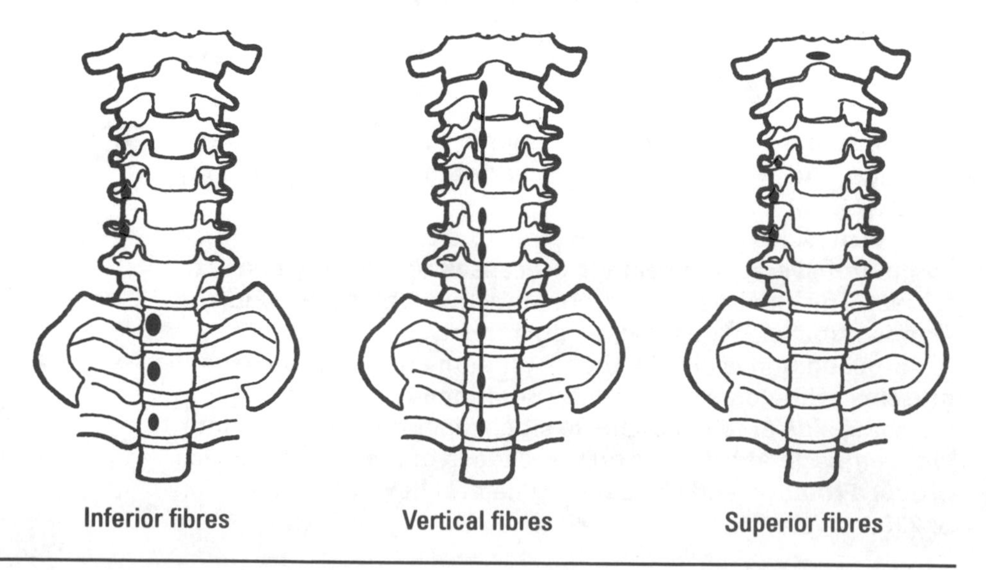

Inferior fibres　　Vertical fibres　　Superior fibres

Attachments

- Inferior fibres:
  anterior vertebral bodies of T1–3;
  anterior tubercles of TPs C5–6.
- Vertical fibres:
  vertebral bodies of C5–T3.
  vertebral bodies of C2–4.
- Superior fibres:
  anterior tubercles of TPs C3–5;
  anterior arch of C1 (atlas).

Nerve supply

Ventral rami of C2–6.

Functions

- Bilaterally, flexion of the neck.
- Unilaterally, sidebending of the neck; the inferior fibres also rotate the neck to the opposite side.

Study task

- Carefully draw lines between the attachment points of each part of the muscle in turn. Consider the individual function of each part as you proceed.

Longus capitis ('long muscle of the head')

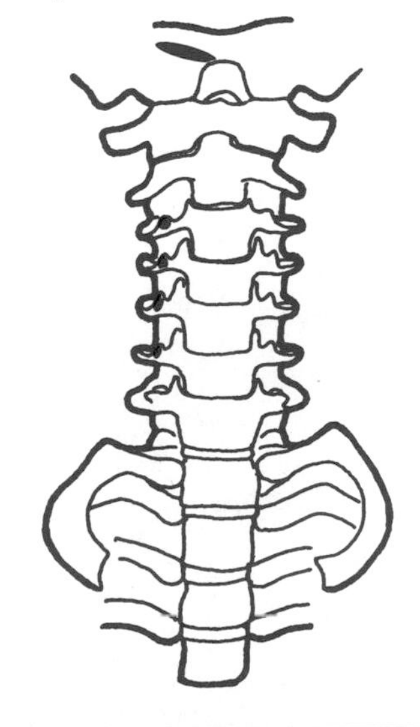

Attachments

- Anterior tubercles of TPs C3–6.
- Inferior surface of basilar occiput.

Nerve supply

Ventral rami of C1–3.

Functions

- Bilaterally, flexion of the head and neck.
- Unilaterally, sidebending of the head and neck.

Study task

- Draw lines between the attachment points and consider the muscle functions.

Rectus capitis anterior ('straight front muscle which holds the head upright')

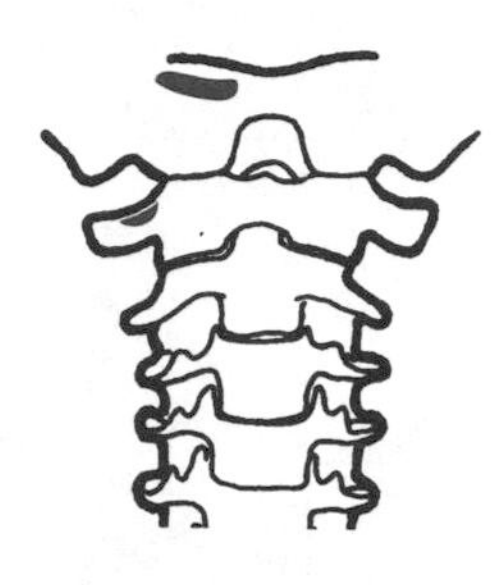

This is a small muscle lying deep to longus capitis.

Attachments

- Anterior surface of the lateral mass and TPs of C1 (atlas).
- Inferior basilar occiput, anterior to the condyle.

Nerve supply

Ventral rami of C1, C2.

Functions

- Bilaterally, flexion of the head.
- Unilaterally, sidebending of the head.

Study task

- Shade lines between the attachment points, and consider the muscle functions.

Flexion of the thoracic and lumbar spine

The thoracic spine is already flexed to a certain extent due to a normal, variable kyphosis. Additional flexion is possible from an upright position under the influence of gravity, controlled by the graded relaxation of the extensor muscles. It occurs actively when sitting up from the supine position.

Flexion of the lumbar spine from an upright position also occurs under the influence of gravity, controlled by the graded relaxation of the vertebral extensor muscles. It occurs actively when sitting up from the supine position, but hip flexion also plays a significant part in the gross movement (pp. 100–101).

(Hip flexion makes angular measurement difficult.
Linear stretch from L5 to L1 is often measured.
Approximate maximum flexion ranges are:
thoracic (0°–45°); lumbar (0°–60°)

- abdominal muscles (pp. 182–186)
- psoas minor (p. 193)

Note: hip flexors not included

Study tasks

- Highlight the names of the muscles.
- Study the details of each muscle with reference to the page number given.

Psoas minor ('smaller loin muscle')

This muscle lies anterior to psoas major, and is absent in c.40% of the population (*Gray's Anatomy*, 1995)

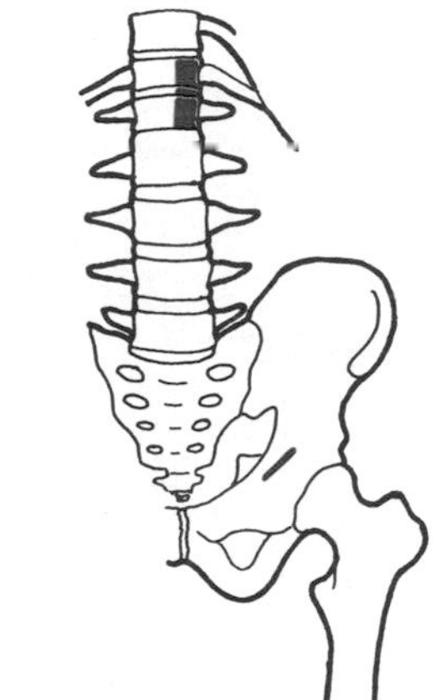

Attachments

- Sides and vertebral bodies of T12 and L1 and the intervening disc.
- Iliopectineal eminence and pecten pubis.

Nerve supply

Branch from L1.

Function

- Weak flexion of the lumbar spine.

The posterior vertebral muscles

Extension of the vertebral column

(Thoracic 0°–25°)
(Lumbar 0°–35°)

(cervical extension may be measured from the horizontal 'plane of bite', (0°–55°)

- trapezius (middle and lower fibres) (pp. 27–28)
- serratus posterior inferior (p. 181)
- erector spinae (pp. 197–203)
- transversospinales (pp. 204–207)
- interspinales (p. 209)
- intertransversarii (p. 208)
- quadratus lumborum (p. 178)

- trapezius (upper fibres) (pp. 27–28)
- splenius capitis (p. 195)
- splenius cervicis (p. 196)
- longissimus capitis and cervicis (p. 200)
- spinalis and semispinalis capitis (p. 198)
- spinalis cervicis (p. 198)
- iliocostalis cervicis (pp. 201–202)
- semispinalis cervicis (p. 206)
- suboccipital muscles (pp. 209–211)

The posterior muscles produce extension of the vertebral column when they act bilaterally, and in practice this often occurs from a position of relative flexion. However, unilateral contraction may produce sidebending (lateral flexion) and rotation, sometimes to the opposite side.

The superficial layer

Readers of this book who are intending to treat patients with musculoskeletal problems will need a clear understanding of the surface anatomy of the back before considering the muscles which lie deeper. The superficial or surface muscles give shape and form to the back, and are readily palpable.

Note also that some muscles, such as deltoid, sternocleidomastoid and the glutei, which have been considered in earlier chapters, are also visible and palpable dorsally.

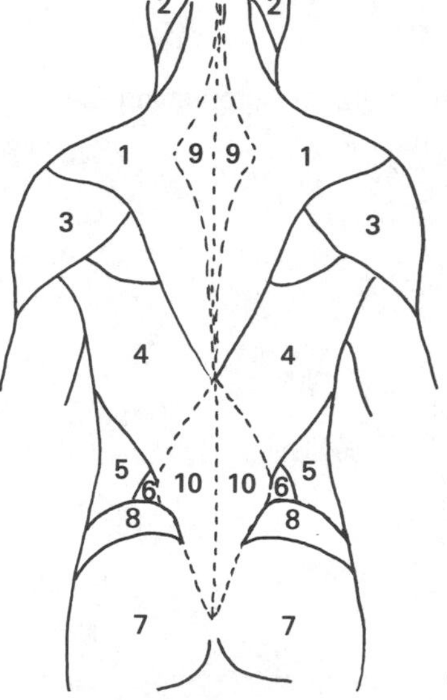

1 Trapezius (pp. 27–28)
2 Sternocleidomastoid (part) (p. 173)
3 Deltoid (p. 32)
4 Latissimus dorsi (p. 37)
5 Obliquus externus abdominis (p. 184)
6 Obliquus internus abdominis (p. 184)
7 Gluteus maximus (p. 108)
8 Gluteus medius (p. 112)
9 Fascia
10 Thoracolumbar fascia

Study tasks

- Shade the muscles in separate colours.
- Highlight the names of the muscles.
- Study the details of each muscle with reference to the page number given.
- Observe and palpate the muscles on a colleague.

The intermediate layer

As the description implies, these muscles lie immediately below the superficial layer, but it is a somewhat artificial classification.

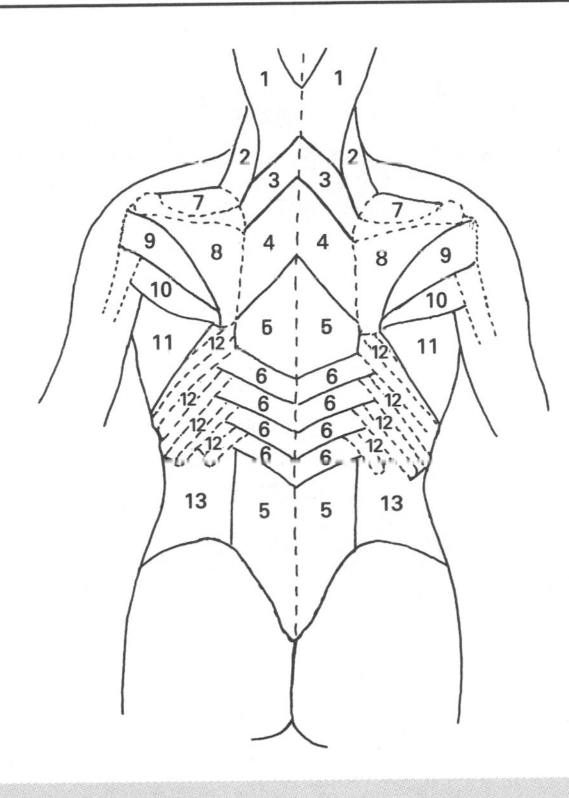

1 Splenius capitis (p. 195)
2 Levator scapulae (p. 26)
3 Rhomboid minor (p. 29)
4 Rhomboid major (p. 29)
5 The underlying erector spinae (pp. 197–203)
6 Serratus posterior inferior (p. 181)
7 Supraspinatus (p. 39)
8 Infraspinatus (p. 42)
9 Teres minor (p. 42)
10 Teres major (p. 36)
11 Serratus anterior (p. 177)
12 Intercostales externi (p. 172)
13 Obliquus internus abdominis (p. 184)

Study tasks

- Shade the muscles in separate colours.
- Highlight the names of the muscles shown.
- Study the details of each muscle with reference to the page number given.

Splenius capitis ('bandage-shaped muscle of the head')

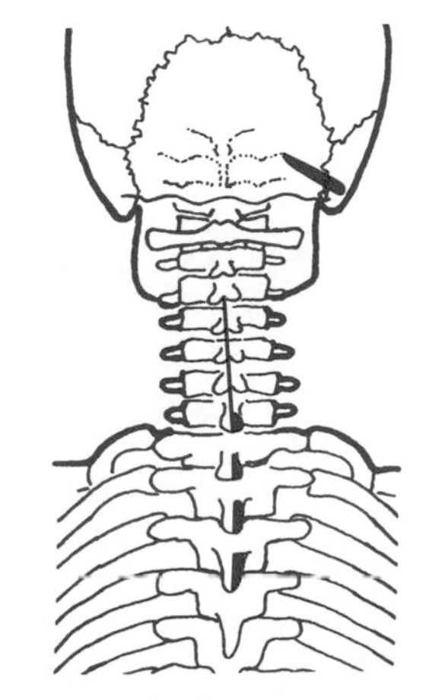

Attachments

- SPs of C7–T3 and lower half of ligamentum nuchae.
- Mastoid process and lateral part of the superior nuchal line.

Nerve supply

Dorsal rami of C3, C4, C5.

Functions

- Bilaterally, extension of the head and neck.
- Unilaterally, sidebending and rotation of head and neck to the same side.

Study task

- Shade lines between the attachment points and consider the muscle functions.

Splenius cervicis ('bandage-shaped muscle of the neck')

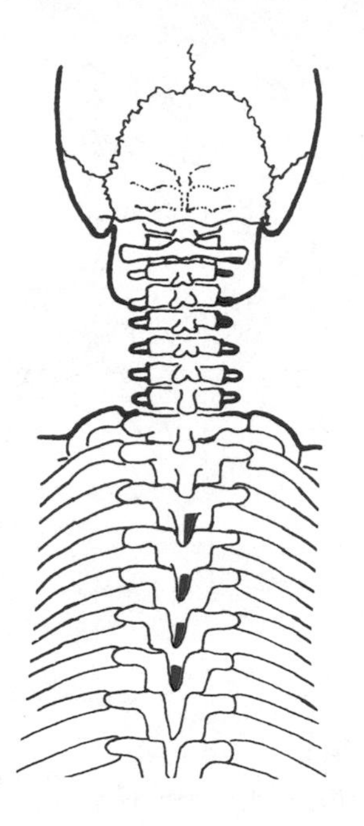

Attachments

- ■ SPs of T3–6.
- ■ TPs of C1–4.

Nerve supply

Dorsal rami of C6, C7, C8.

Functions

- ♦ Bilaterally, extension of the head and neck.
- ♦ Unilaterally, sidebending and rotation of head and neck to the same side.

Study task

- Shade lines between the attachment points and consider the muscle functions.

The deep layer (erector spinae)

Below the intermediate layer lie the erector spinae (sacrospinalis) muscles in three medial–lateral 'bands', with each band variously represented according to whether it has attachment to the head (capitis), cervical spine (cervicis), thoracic spine (thoracis) or lumbar spine (lumborum). These details may be summarized in a table:

Distribution of the erector spinae (sacrospinalis) muscles

Region	Spinalis	Longissimus	Iliocostalis
Capitis	✓	✓	
Cervicis	✓	✓	✓
Thoracis	✓	✓	✓
Lumborum			✓

Superior ↕ Inferior

Medial ↔ Lateral

Study tasks

- Highlight the names of the muscles shown on the chart.
- Observe and palpate the erector spinae muscles functioning as a group on a healthy colleague. They should lie prone and then raise their head and straightened legs approximately 1 cm only. The muscles will be seen and felt to contract on both sides of the spinous processes, which form a furrow.

The spinalis muscles ('muscles close to the spinous processes')

This is the most medial, yet curiously indistinct part of the erector spinae group. Spinalis cervicis is often not present at all, and spinalis capitis and thoracis tend to blend with other muscles (*Gray's Anatomy* 1995).

(i) Spinalis capitis/semispinalis capitis ('muscles of the head which are close, and fairly close to the spinous processes')

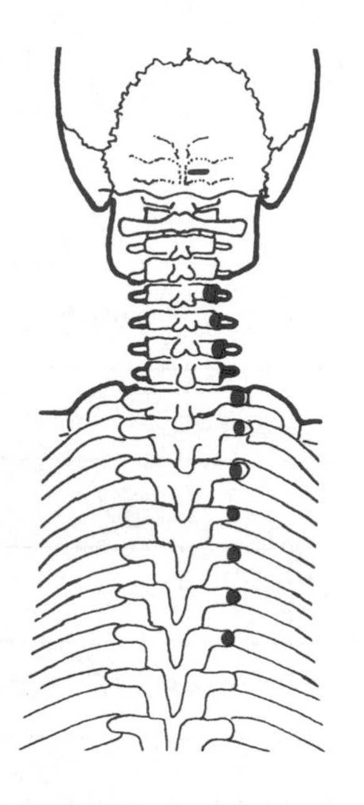

These two muscles blend and will therefore be considered together.

Attachments

- TPs of C7, T1–7 and articular processes of C4–6.
- Between the superior and inferior nuchal lines of the occiput.

Nerve supply

Dorsal rami of adjacent segmental nerves.

Functions

- Bilaterally, extension of head and neck.
- Unilaterally, sidebending and slight rotation to the opposite side.

> **Study task**
>
> • Shade lines between the attachment points and consider the muscle functions.

(ii) Spinalis cervicis ('muscle of the neck, close to the spinous process')

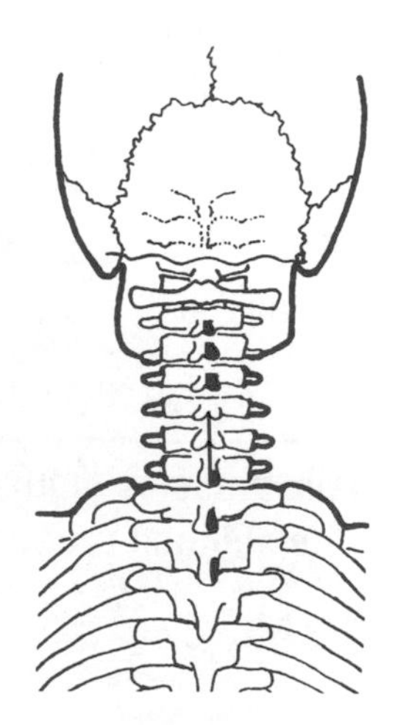

Attachments

- SPs of C7, T1–2 and lower part of ligamentum nuchae.
- SPs of C2–4.

Nerve supply

Dorsal rami of adjacent segmental nerves.

Functions

- Bilaterally, extension of the neck.
- Unilaterally, sidebending of the neck.

> **Study task**
>
> • Shade lines between the attachment points and consider the muscle functions.

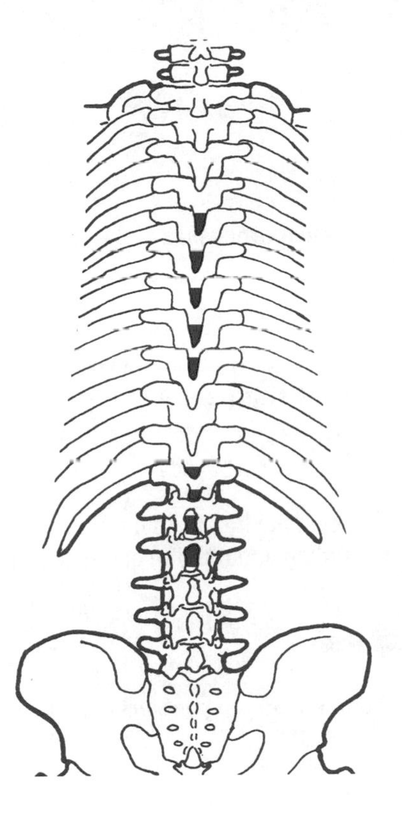

(iii) Spinalis thoracis ('muscle of the thorax, close to the spinous processes')

Attachments

- SPs of T11, T12, L1–2.
- SPs of T4–8.

Nerve supply

Dorsal rami of adjacent segmental nerves.

Function

- Bilaterally, extension of the thoracic spine.

Study task

- Shade lines between the attachment points and consider the muscle functions.

The longissimus muscles ('longest muscles')

(i) Longissimus capitis ('longest muscles of the head')

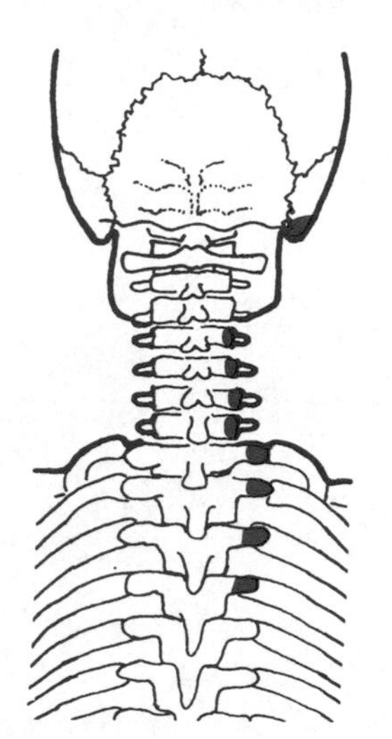

Attachments

- TPs of T1–5 and articular processes of C4–7.
- Mastoid process.

Nerve supply

Dorsal rami of adjacent segmental nerves.

Functions

- Bilaterally, extension of the head and neck.
- Unilaterally, sidebending and rotation of the head to the same side.

Study task

- Shade lines between the attachment points and consider the muscle functions.

(ii) Longissimus cervicis ('longest muscles of the neck')

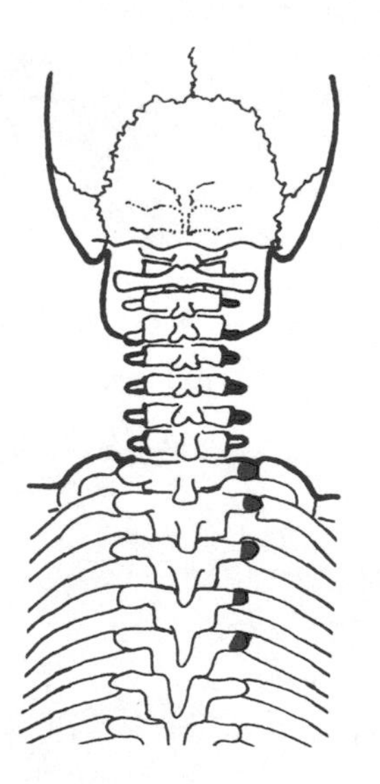

Attachments

- TPs of T1–5.
- TPs of C2–6.

Nerve supply

Dorsal rami of adjacent segmental nerves.

Functions

- Bilaterally, extension of the neck.
- Unilaterally, sidebending of the neck.

Study task

- Shade lines between the attachment points and consider the muscle functions.

(iii) Longissimus thoracis ('longest muscle of the thorax')

The largest of the erector spinae, this muscle originates with iliocostalis lumborum (p. 203) on the medial iliac crest.

Attachments

- Medial iliac crest.
- TPs and accessory processes of L1–5.
- Thoracolumbar fascia (not shown).
- TPs of T1–12.
- Ribs 3–12 between the tubercles and angles.

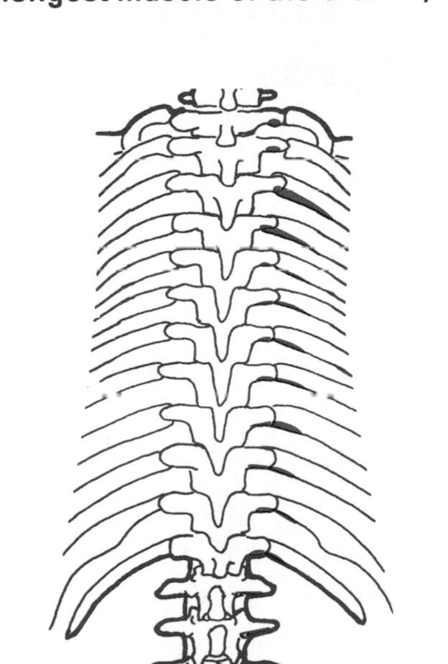

Nerve supply

Dorsal rami of adjacent segmental nerves.

Functions

- Bilaterally, extension of the thoracic spine.
- Unilaterally, sidebending of the thoracic spine.
- As an accessory muscle of respiration, it may draw the ribs downwards in expiration.

Study task

- Shade lines between the attachment points and consider the muscle functions.

The iliocostalis (iliocostocervicalis) muscles ('muscles with iliac/rib/neck attachments')

(i) Iliocostalis cervicis ('muscle of the neck with iliac and rib attachments')

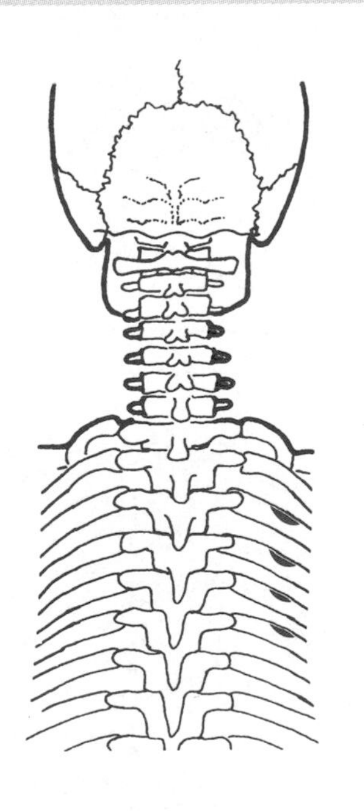

Attachments

- Angles of ribs 3–6.
- TPs of C4–6.

Nerve supply

Dorsal rami of adjacent segmental nerves.

Functions

- Bilaterally, extension of lower cervical and upper thoracic spine.
- Unilaterally, sidebending of the same area.
- As an accessory muscle of respiration, it draws the ribs upwards in inspiration (Kapandji, 1987).

Study task

- Shade lines between the attachment points and consider the muscle functions.

(ii) Iliocostalis thoracis ('muscle of the thorax with iliac and rib attachments')

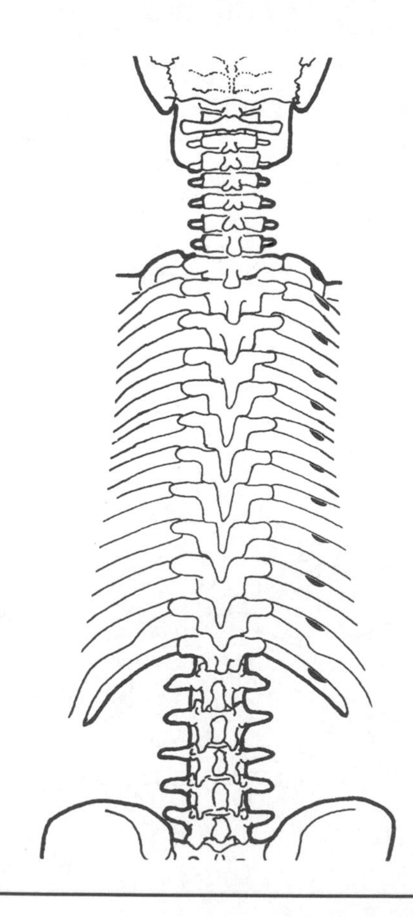

Attachments

- Angles of ribs 7–12 (upper borders).
- Angles of ribs 1–6 (upper borders) and TP of C7.

Nerve supply

Dorsal rami of adjacent segmental nerves.

Functions

- Bilaterally, extension of the thoracic spine.
- Unilaterally, sidebending of the thoracic spine.

Study task

- Shade lines between the attachment points and consider the muscle functions.

(iii) Iliocostalis lumborum ('muscle of the low back with iliac and rib attachments')

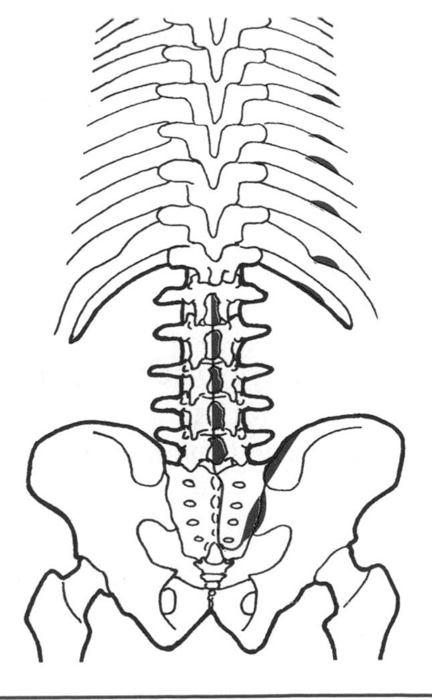

With longissimus thoracis, iliocostalis lumborum forms the 'root' of the erector spinae. The SP attachments are indistinct, blending with the fascia.

Attachments

- Median sacral crest.
- Lateral sacrum blending with the dorsal sacro-iliac and sacrotuberous ligaments, and muscle fibres of gluteus maximus.
- SPs of L1–L5 and supraspinous ligament (fascial attachments).
- Medial aspect of the iliac crest.
- Inferior border of angles of ribs 7–12.

Nerve supply

Dorsal rami of adjacent segmental nerves.

Functions

- Bilaterally, extension of the thoracic and lumbar spine.
- Unilaterally, sidebending of the same area.
- As an accessory muscle of respiration it may draw the ribs downwards in expiration.

Study task

- Shade lines between the attachment points and consider the muscle functions.

The deepest intrinsic layer

These muscles have an important postural function as well as producing extension, rotation and sidebending. They lie deep in the bony 'gutter' between the transverse and spinous processes of the vertebrae (transversospinales), and in the gaps between them (intertransversarii and interspinales).

Transverse section and superior view of a typical thoracic vertebra to show the transversospinalis arrangement of muscles. (Schematic representation.)

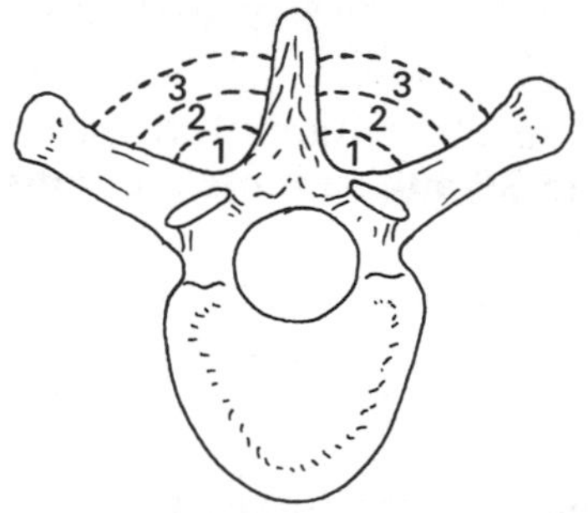

1 Rotatores (pp. 204–205)
2 Multifidus (p. 205)
3 Semispinalis (pp. 206–207)

Study tasks

- Shade the muscles shown in separate colours.
- Highlight the names of the muscles.
- Study the details of each muscle with reference to the page number given.

The transversospinalis group ('muscles attached between transverse and spinous processes')

These muscles pass obliquely upwards and medially, to give rotation to the opposite side when acting unilaterally. Bilaterally, they maintain erect posture of the spine whether standing or sitting.

(i) Rotatores ('rotator muscles')

These muscles are found throughout the spine, and are designated rotatores cervicis, thoracis and lumborum, but are best developed in the thoracic region, and are illustrated here using a thoracic segment to show their *modus operandi*.

Attachments

- Upper and posterior part of TP below.
- Lower and lateral part of lamina of vertebra above.

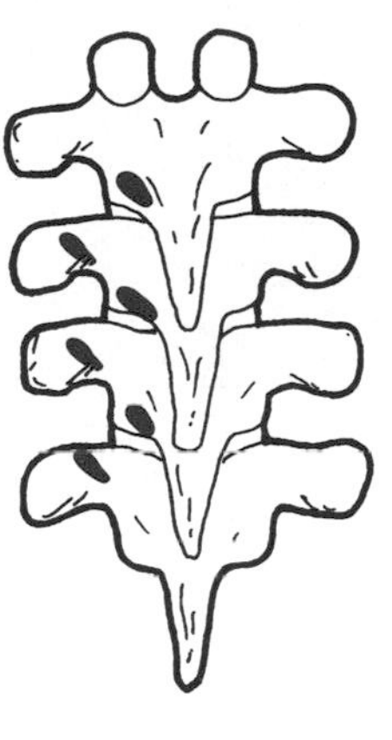

Nerve supply

Dorsal rami of adjacent segmental nerves.

Functions

- Bilaterally, segmental extension of the vertebral column.
- Bilaterally, maintenance of erect spinal posture.
- Unilaterally, rotation to the opposite side.

Study task

- Shade lines between the attachment points on the segmental sample given, and consider the functions of the muscle throughout the entire spine.

(ii) Multifidus ('many-stranded muscle')

The complex intricacy of this muscle (really a series of small muscles), stretching from the sacrum to C2, make the adopted practice in this book of shading lines between attachment points impractical. As with rotatores, a sample segment is given to illustrate the principles of action (in the lumbar spine).

Attachments

- The basic principle is that each muscle passes from a relatively lateral point (such as a TP) on a vertebra below, obliquely upwards to a more medial attachment (such as an SP) on a vertebra above. The lateral attachment points are variously:
  intermediate dorsum of the sacrum;
  aponeurosis of the erector spinae;
  posterior superior iliac spine;
  dorsal sacro-iliac ligament;
  lumbar mamillary processes;
  TPs of thoracic vertebrae (illustrated);
  articular processes of C4–7
- The deepest fasciculi then pass to the SP of the vertebra immediately above.
- The intermediate fasciculi pass to the SP of the second or third vertebra above.
- The most superficial fasciculi pass to the SP of the third or fourth vertebra above.

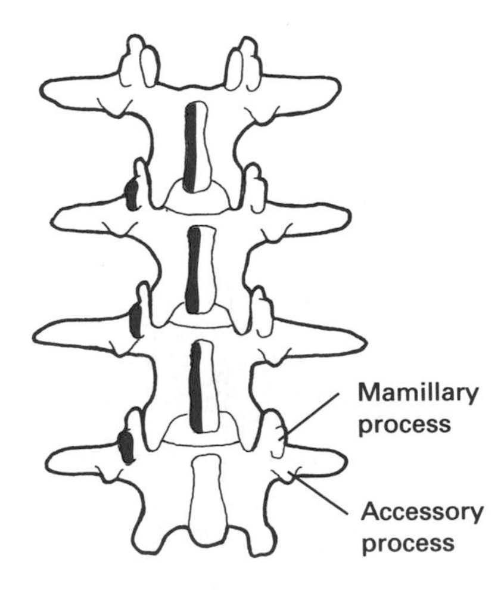

The arrangement of the deepest fasciculi

Nerve supply

Dorsal rami of adjacent segmental nerves.

Functions

- Bilaterally, extension of the vertebral column.
- Bilaterally, maintenance of erect spinal posture.
- Unilaterally, slight sidebending with rotation to the opposite side.

Study task

- Shade lines between the attachment points of the sample shown on each of the three layers and consider the muscle functions.

(iii) Semispinalis ('one-half attached to the spinous processes')

These are the most superficial of the deep intrinsic muscles, and broaden in the cervical spine to form a muscular layer just beneath splenius capitis. Nevertheless, they form part of the transversospinalis group, and conform to the general principle of attachment from TP below to SP above, which produces rotation to the opposite side when the muscle contracts unilaterally.

Semispinalis capitis

This muscle blends with spinalis capitis (see under **spinalis capitis**, p. 198).

Semispinalis cervicis ('semispinalis muscle of the neck')

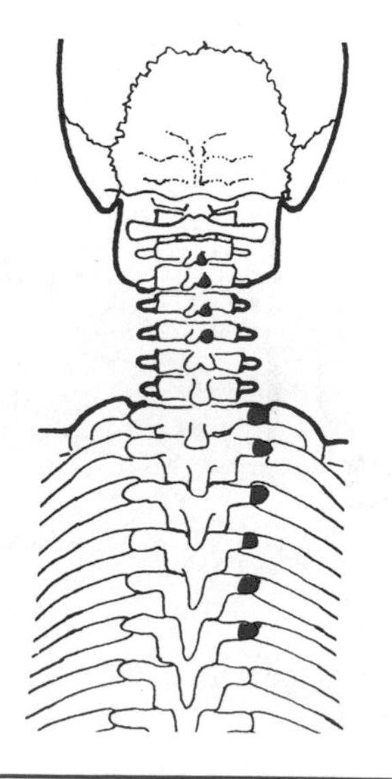

Attachments

- TPs of T1–6.
- SPs of C2–5.

Nerve supply

Dorsal rami of adjacent segmental nerves.

Functions

- Bilaterally, extension of the cervical spine.
- Unilaterally, rotation of the cervical spine to the opposite side.

Study task

- Shade lines between the attachment points and consider the muscle functions.

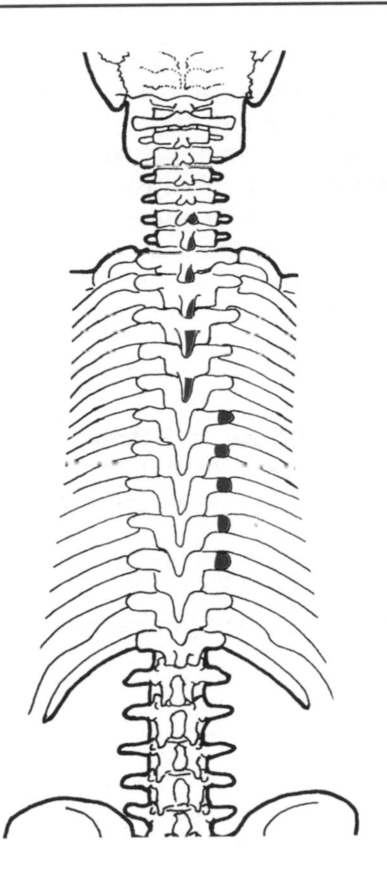

Semispinalis thoracis ('semi-spinalis muscle of the thorax')

Attachments

- ■ TPs of T6–10.
- ■ SPs of C6–T4.

Nerve supply

Dorsal rami of adjacent segmental nerves.

Functions

- ♦ Bilaterally, extension of the thoracic spine.
- ♦ Bilaterally, maintenance of erect spinal posture.
- ♦ Unilaterally, rotation of the thoracic spine to the opposite side.

Study task

- Shade lines between the attachment points and consider the muscle functions.

The intertransversarii group ('muscles between the transverse processes')

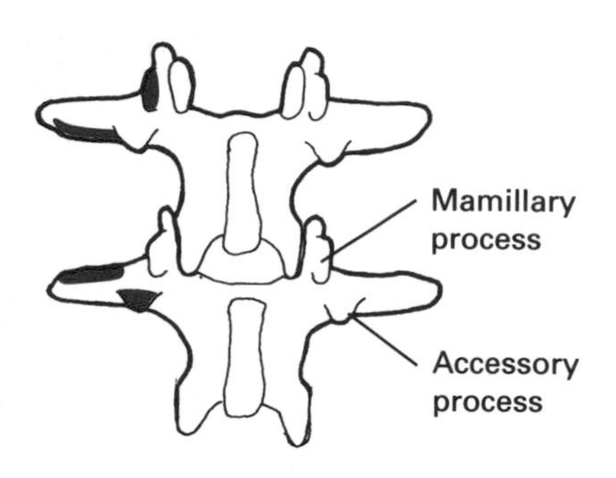

These small intrinsic muscles link the transverse processes of adjacent vertebrae, adding stability bilaterally and sidebending individual segments unilaterally. They are well developed in the lumbar spine (shown here) and in the cervical spine, where they divide into anterior and posterior slips separated by the ventral rami of the cervical nerves.

In the thoracic spine, they consist of single muscles present between T10 and L1 only. Elsewhere in the thoracic spine, they are replaced in effect by the intertransverse ligaments.

Attachments (lumbar)

- Lateral band from TP below to TP above.
- Medial band from the accessory process of the vertebra below to the mamillary process of the vertebra above.

Nerve supply

Variously by the dorsal and ventral rami of the adjacent segmental nerves.

Functions

- Bilaterally, extension of the vertebral column.
- Bilaterally, maintenance of erect spinal posture.
- Unilaterally, sidebending of the vertebral column.

Study task

- Shade lines between the attachment points of the lumbar intertransversarii and consider the muscle functions.

Rectus capitis lateralis ('outer muscle which holds the head upright')

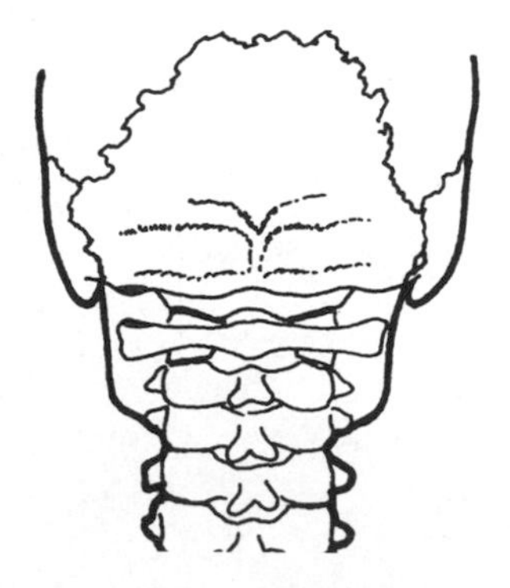

This small muscle is also generically part of the intertransversarii group and shares the same general functions.

Attachments

- TP of the atlas.
- Jugular process of the occiput.

Nerve supply

Ventral rami of C1–2.

Functions

- Bilaterally, postural maintenance of an erect head.
- Unilaterally, sidebending of the head.

Study task

- Shade lines between the attachment points and consider the muscle functions.

The interspinales group ('muscles between the spinous processes')

These are short, paired muscles which connect the spinous processes of the vertebrae and lie each side of the interspinous ligament. They are deficient in the mid-thoracic area.

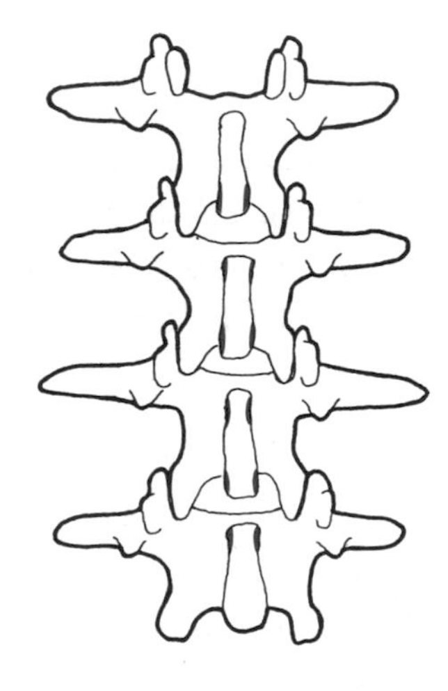

Attachments

- SP to SP.

Nerve supply

Dorsal rami of adjacent segmental nerves.

Function

- Postural stability and contribution to extension of the vertebral column.

Study task

- Shade lines between the attachment points and consider the muscle functions.

The suboccipital group ('muscles just below the occiput')

The suboccipital muscles are a small group of deep muscles which extend, sidebend and to some extent rotate the head and atlas on the axis (C2).

(i) Rectus capitis posterior major ('larger muscle at the back which holds the head upright')

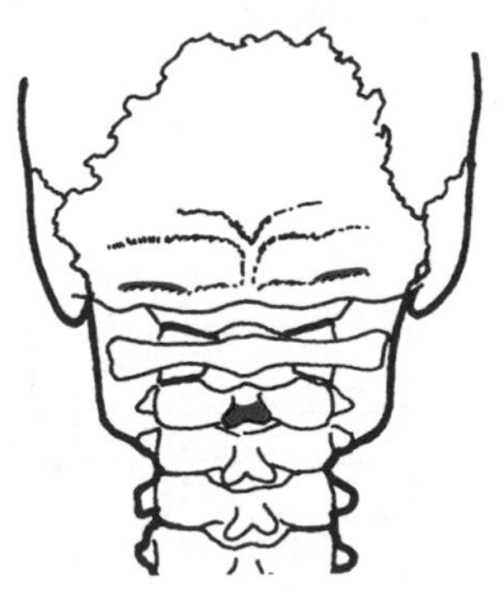

Attachments

- Spinous process of the axis.
- Lateral aspect of the inferior nuchal line of the occiput.

Nerve supply

Dorsal ramus of C1 spinal nerve.

Functions

- ♦ Bilaterally, extension of the head.
- ♦ Bilaterally, postural maintenance of an erect head.
- ♦ Unilaterally, rotation of the head to the same side, with some sidebending.

Study task

- Shade lines between the attachment points and consider the muscle functions.

(ii) Rectus capitis posterior minor ('smaller muscle at the back which holds the head upright')

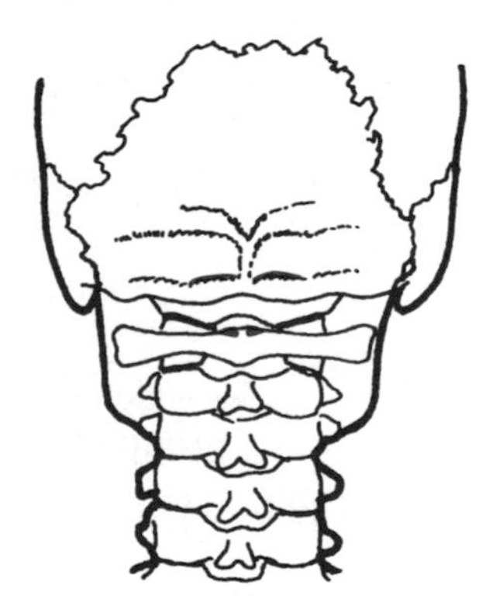

Attachments

- ■ Tubercle on the posterior arch of the atlas.
- ■ Medial aspect of the inferior nuchal line of the occiput.

Nerve supply

Dorsal ramus of C1 spinal nerve.

Functions

- ♦ Bilaterally, extension of the head.
- ♦ Bilaterally, postural maintenance of an erect head.

Study task

- Shade lines between the attachment points and consider the muscle functions.

(iii) Obliquus capitis superior ('upper angled muscle of the head')

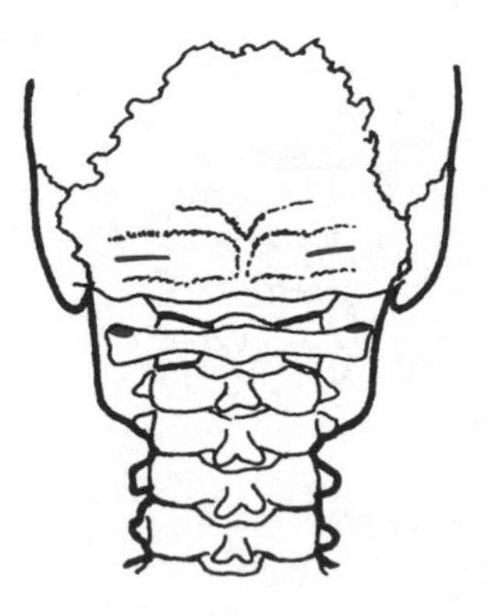

Attachments

- TP of the atlas.
- Insertion between the inferior and superior nuchal lines overlapping rectus capitis posterior major.

Nerve supply

Dorsal ramus of C1 spinal nerve.

Functions

- Bilaterally, extension of the head.
- Bilaterally, postural maintenance of an erect head.
- Unilaterally, slight sidebending of the head.

Study task

- Shade lines between the attachment points and consider the muscle functions.

(iv) Obliquus capitis inferior ('lower angled muscle of the head')

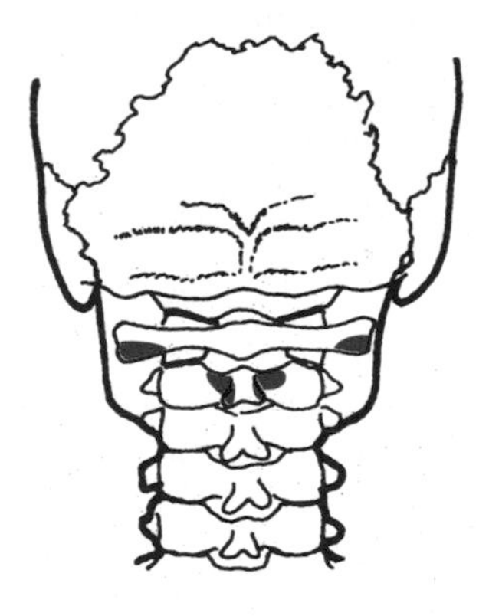

Attachments

- Spinous process and lamina of the axis.
- TP of the atlas.

Nerve supply

Dorsal ramus of C1 spinal nerve.

Functions

- Unilaterally only, rotation of the head to the same side.

(Bilateral contraction of this muscle, as well as hypertonic contraction of the other suboccipital muscles, probably contributes to the commonly experienced 'tension headache'. Proximity to the foramen magnum may also produce a headache due to strain on the dura mater.)

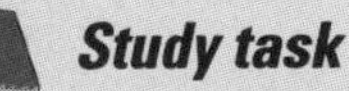

Study task

- Shade lines between the attachment points and consider the muscle functions.

Rotation and sidebending (lateral flexion)

Sidebending (lateral flexion) of the cervical spine (0°– 45°) (this may be measured from the horizontal 'plane of bite')

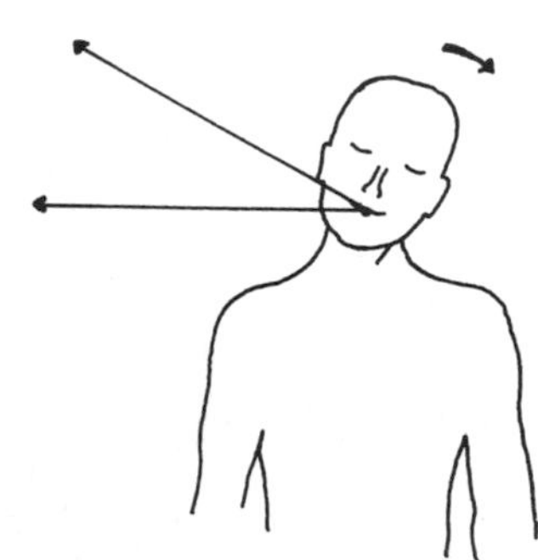

Contraction of muscles on the same side:

- longus colli (pp. 90–91)
- longus capitis (p. 191)
- rectus capitis lateralis (p. 208)
- suboccipital muscles (pp. 209–211)
- sternocleidomastoid (p. 173)
- trapezius (upper fibres) (pp. 27–28)
- spinalis cervicis (p. 198)
- splenius capitis and cervicis (pp. 195 and 198)
- spinalis capitis/semispinalis capitis (p. 198)
- longissimus capitis and cervicis (p. 200)
- levator scapulae (p. 26)
- scalenes (pp. 170–172)
- intertransversarii (p. 208)

Sidebending (lateral flexion) of the thoracic and lumbar spine (0°–30°)

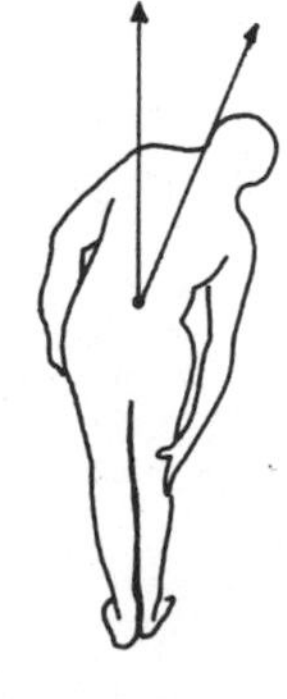

Contraction of muscles on the same side:

- longissimus muscles (pp. 200–201)
- iliocostalis muscles (pp. 201–204)
- abdominal oblique muscles (pp. 182–186)
- psoas major (p. 102)
- quadratus lumborum (p. 178)
- intertransversarii (p. 208)
- levatores costarum (p. 169)

Rotation of the cervical spine (0°– 80° on either side)

Contraction of muscles on the same side:

- obliquus capitis inferior (p. 211)
- rectus capitis posterior major (pp. 209–210)
- splenius capitis and cervicis (pp. 195, 196)
- longissimus capitis (p. 200)

Contraction of muscles on the opposite side:

- sternocleidomastoid (p. 173)
- scalenus anterior (p. 170)
- longus colli (inferior fibres) (pp. 190–191)
- semispinalis capitis and cervicis (pp. 198 and 206)
- rotatores (pp. 204–205)
- multifidus (p. 205)
- trapezius (upper fibres) (pp. 27–28)

Rotation of the thoracic and lumbar spine (thoracic: 0°– 30°; lumbar: 0°– 5°)

Contraction of muscles on the same side:

- obliquus abdominis internus (p. 184)
- levatores costarum (p. 169)

Contraction of muscles on the opposite side:

- rotatores (pp. 204–205)
- multifidus (p. 205)
- semispinalis thoracis (p. 207)
- obliquus abdominis externus (p. 184)

Study tasks

- Highlight the names of the features indicated, including the names of the muscles.
- Study the details of each muscle with reference to the page number given.

The orientation of the apophyseal facet joints (facets) allows fairly free rotation in the cervical spine, with between one-third and one-half of the movement occurring at the atlanto-axial joint (C1/C2). Rotation is also the main movement permitted in the thoracic spine, since the coronal plane of the facets allows rotation, and the ribs are relatively less restrictive in this movement. In contrast, the sagittal plane of the lumbar facets, as well as the influence of the pelvis, do not encourage much rotation in the lumbar spine.

To some extent, sidebending or lateral flexion accompanies rotation quite naturally (and *vice versa*), but increased sidebending is possible with the unilateral recruitment of some of the muscles already described. Once again, the orientation of the facets plays an important part. In the cervical spine the range of movement is also limited by the lateral lipping (uncinate/unciform processes) of the vertebral bodies. In the thoracic spine, the ribs inhibit movement; and in the lumbar spine, the sagittal plane of the facets and the relative lack of movement in the pelvis are both restrictive influences.

Chapter 14

The temporomandibular joint

Introduction

The temporomandibular joint is part of that complex system which includes the cranial bones, teeth, hyoid bone and the soft tissues of head, neck and face.

It is therefore part of a highly specialized area which is involved not only with the functions of eating and swallowing, respiration and speech, but also with more subtle considerations of posture, balance and feelings of well-being.

It is only within the scope of this book to provide a basic outline of the muscles acting mainly on the temporomandibular joint itself, and more specialized texts should be consulted as appropriate.

The bones

① Temporal bone
② Mandible
③ Sphenoid bone (greater wing)
④ Zygomatic bone
⑤ Maxilla
⑥ Frontal bone
⑦ Parietal bone
⑧ Occipital bone

Selected bone features:

6 External occipital protruberance
7 Mastoid process
8 External auditory meatus
9 Styloid process
10 Zygomatic arch of temporal bone
11 Head (of mandible)
12 Neck (of mandible)
13 Coronoid process (of mandible)
14 Ramus (of mandible)
15 Angle (of mandible)
16 Body (of mandible)
17 Temporal lines (surrounding temporal fossa)

Study tasks

- Shade the bones shown in separate colours.
- Highlight the names of the bone features shown.

The joint and ligaments

Temporomandibular joint (condylar with articular disc)

Lateral view (right)

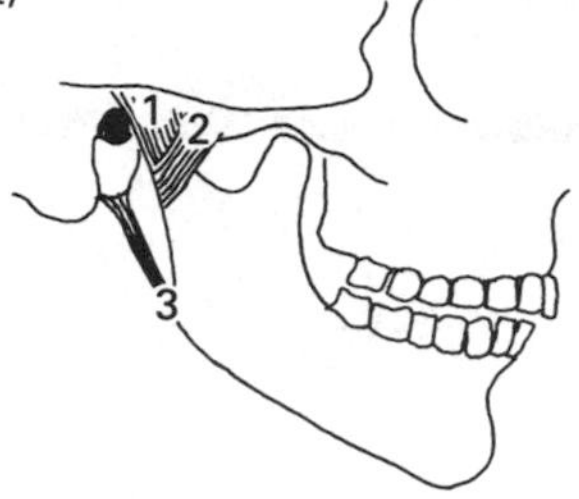

Medial view (left)

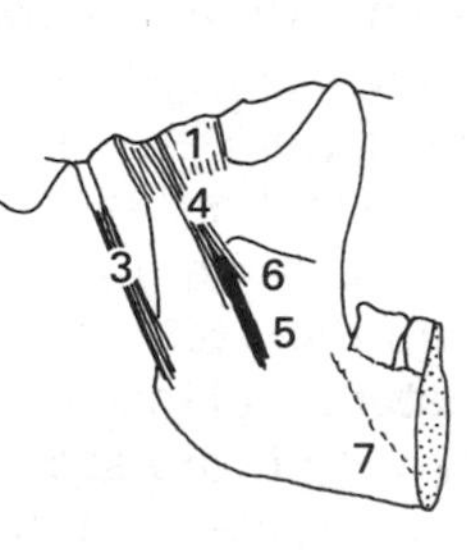

1 Capsule
2 Lateral ligament
3 Stylomandibular ligament
4 Sphenomandibular ligament
5 Mylohyoid groove
6 Lingula
7 Mylohyoid line (part)

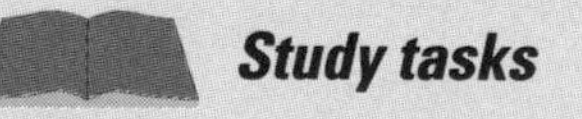

Study tasks

- Shade the ligaments in colour.
- Highlight the names of the features shown.

Movements

Depression (opening)

The mandibular condyles slide forwards and downwards with the articular discs, allowing the jaw to open. The movement continues until the fibro-elastic disc is stretched to its limit of comfort. Gravity plays some part, but the movement is effected by the lower part of the lateral pterygoid muscle, while the anterior part of the digastric and the geniohyoid muscles draw the mandible toward the bony platform of the hyoid. The hyoid itself is held steady by the synergistic contraction of the infrahyoid muscles (from below) and the suprahyoid muscles (from above).

Depression of the mandible

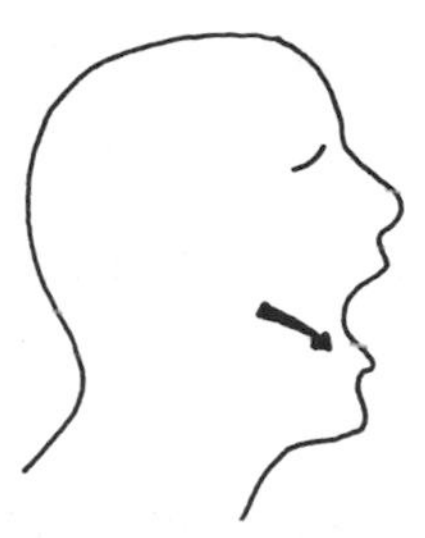

- lateral pterygoid (p. 217)

assisted by:

- gravity
- digastric (p. 218)
- mylohyoid (pp. 218–219)
- geniohyoid (p. 219)

Study tasks

- Highlight the names of the muscles shown.
- Study the details of each muscle with reference to the page numbers given.

Lateral pterygoid ('outer wing-like muscles')

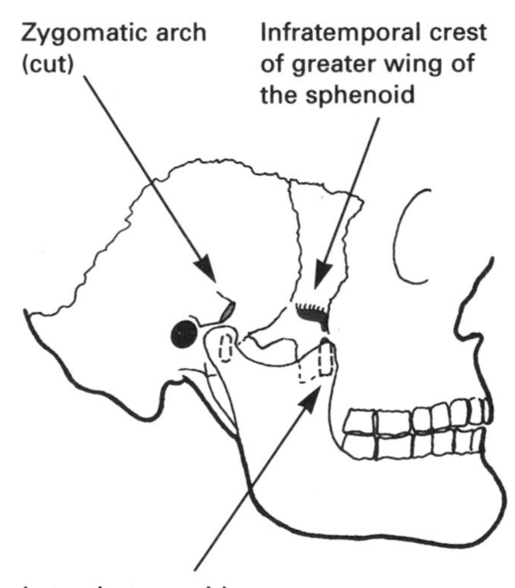

Attachments

- Upper head from the greater wing of the sphenoid bone below the infratemporal crest.
- Lower head from the lateral side of the lateral pterygoid plate.
- Common tendinous insertion into front of neck of the mandible and articular disc.

Nerve supply

Anterior branch of mandibular division of cranial nerve V (trigeminal).

Functions

- Depression of the mandible is achieved by drawing the articular disc and condylar heads forward. The lower head may be more active in this movement.
- Protrusion of the mandible, which is a deliberate act of drawing the mandible forwards and is also effected by the lower head.
- Unilateral protrusion (one side only) in mastication (chewing) (p. 224).
- Stability of the temporomandibular joint during closure of the mandible in mastication is achieved by the upper head.

Study task

- Shade lines between the attachment points of both heads of the muscle, and consider their functions.

The suprahyoid muscles ('the muscles attached to, lying above, the hyoid bone')

(i) Digastric ('two-bellied muscle')

As the name implies, there are two parts to this muscle, united by a common fibrous loop on the hyoid bone.

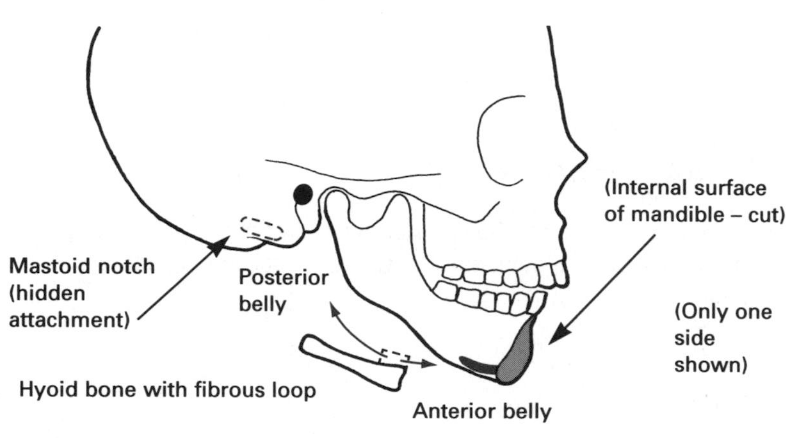

Attachments

- The posterior belly from the mastoid notch of the temporal bone.
- The anterior belly from the base of the mandible, close to the midline.
- Both bellies are united by an intermediate tendon which is connected to the hyoid bone by a fibrous loop.

Nerve supply

Mylohyoid branch of the inferior alveolar nerve (anterior belly); facial nerve (cranial nerve VII) (posterior belly).

Function

- Depression of the mandible with elevation of the hyoid bone, especially during maximum use.

Study task

- Draw lines between the attachment points and consider the muscle functions.

(ii) Mylohyoid ('molar', or literally 'muscle of the "mill" and hyoid bone')

This muscle (the derivation of its name is slightly puzzling), lies deep to the anterior digastric, and helps form a muscular floor for the mouth.

Attachments

- The mylohyoid line on the inner surface of the mandible.
- The hyoid bone.

(The fibres from each side intersect at a midline raphe.)

Nerve supply

Mylohyoid branch of the inferior alveolar nerve.

Functions

- Depression of the mandible if the hyoid is fixed by the infrahyoid muscles.
- Elevation of the hyoid bone and the floor of the mouth (when swallowing) if the mandible is fixed.

Supero-posterior view of mandible and hyoid

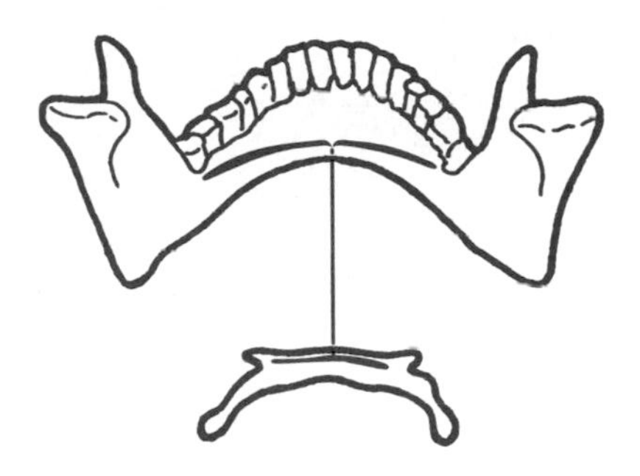

Study task

- Shade lines between the attachment points and consider the muscle functions.

(iii) Geniohyoid ('muscle of the chin and hyoid bone')

A narrow muscle which has virtually the same action as mylohyoid and lies above it, each side of the midline.

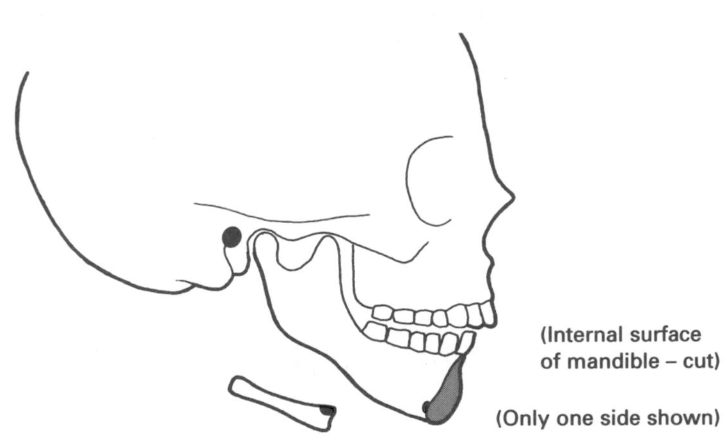

Attachments

- Posterior symphysis menti.
- Anterior surface of the hyoid bone.

Nerve supply

C1 through the hypoglossal nerve.

Functions

- As for mylohyoid above.

Study task

- Shade lines between the attachment points and consider the muscle functions.

The remaining suprahyoid muscle, **stylohyoid**, passing from the styloid process to the hyoid bone, elevates and draws back the hyoid, and therefore plays a role in swallowing and speech processes. In the contexts described here, its main function would appear to be synergistic, holding the hyoid steady. The **infrahyoid muscles**, the muscles attached to, but below the hyoid bone (**thyrohyoid, omohyoid, sternohyoid** and **sternothyroid**), also act as fixators of the hyoid bone from below, while the suprahyoid muscles act from above.

Elevation

The articular discs and condylar heads glide back helped by the graded relaxation of the lateral pterygoids. The upper part of the lateral pterygoid may help to guide the articulation before the final powerful closure (occlusion) of the mandible is carried out by masseter, medial pterygoid and temporalis muscles.

Elevation (closure) of the mandible

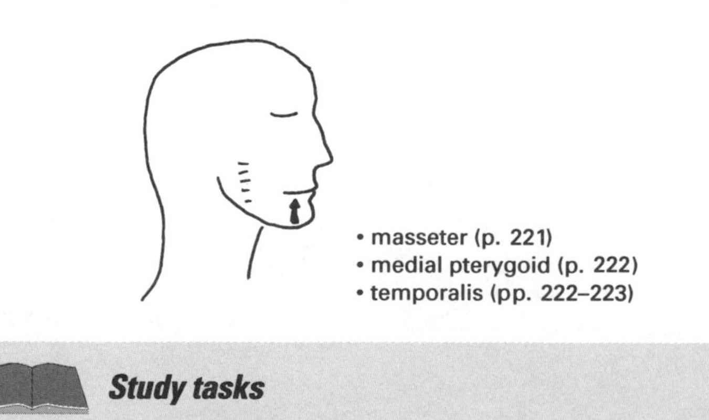

- masseter (p. 221)
- medial pterygoid (p. 222)
- temporalis (pp. 222–223)

Study tasks

- Highlight the names of the muscles.
- Study the details of each muscle with reference to the page number given.

Masseter ('muscle of chewing')

This quadrilateral muscle consists of three layers which blend to form the most powerful muscle in the body. The superficial and middle fibres are angled to allow some protrusion while the deep fibres produce a degree of retraction. Together they produce an extremely powerful 'bite'.

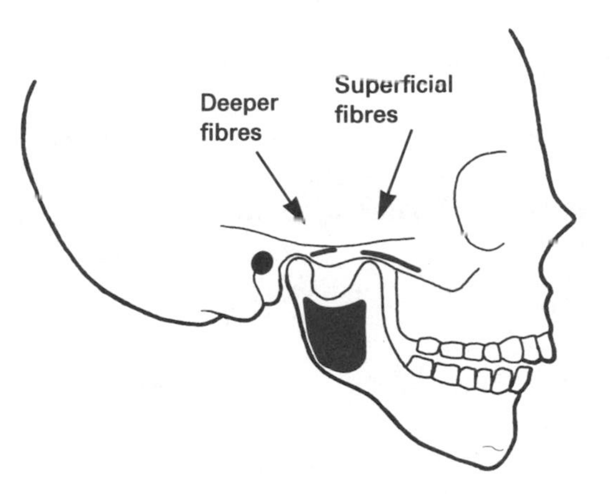

Attachments

- Superficial fibres from the zygomatic process and border of the zygomatic arch.
- Deeper fibres from the deep surface of the zygomatic arch.
- Superficial fibres insert into the angle and lower lateral ramus of the mandible.
- Deeper fibres insert into the upper ramus and coronoid process.

Nerve supply

Anterior branch of the mandibular division of cranial nerve V (trigeminal).

Functions

- Powerful occlusion of the jaw in mastication.
- The difference in emphasis of fibre direction between superficial and deep fibres allows the alternating protrusion and retraction needed in the rotatory action of mastication.

Study tasks

- Shade lines between the attachment points and consider the muscle functions.
- Palpate the muscle on yourself by gently occluding your jaw, and then contracting masseter more powerfully.

Medial pterygoid ('inner wing-like muscle')

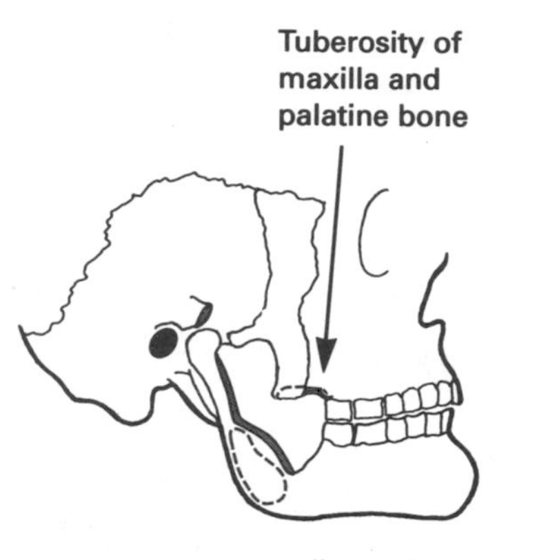

Depicts hidden attachments

(Lateral view of right mandible – cut)

To some extent this muscle can be regarded as the inner equivalent of masseter. It has a similar shape, and the angle of the fibres produces protrusion like the superficial fibres of masseter.

Attachments

- Superficial fibres from the tuberosity of the maxilla and the palatine bone.
- Deeper fibres from the medial side of the lateral pterygoid plate and the palatine bone.
- The fibres merge to insert into the medial ramus and angle of the mandible.

Nerve supply

Branch of the mandibular division of cranial nerve V (trigeminal).

Functions

- Elevation of the mandible.
- Protrusion (with lateral pterygoid and superficial masseter).

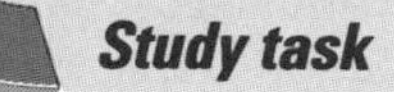

Study task

- Shade lines between the attachment points and consider the muscle functions.

Temporalis ('muscle of time')

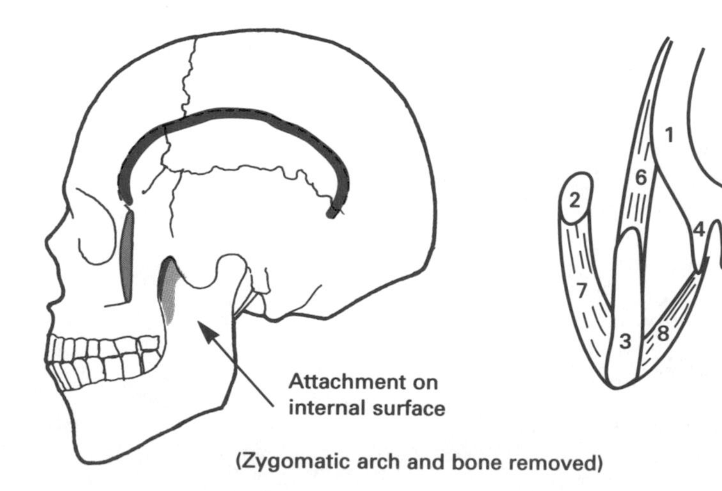

(Zygomatic arch and bone removed)

Schematic coronal section to show muscle attachment

1 Temporal bone
2 Zygomatic arch
3 Mandible
4 Lateral pterygoid plate } of sphenoid
5 Medial pterygoid plate } of sphenoid
6 Temporalis
7 Masseter
8 Medial pterygoid (part)

Note: The pterygoid plates are not part of the temporal bone.

The name derives from the fact that this muscle (together with the temporal fascia) attaches to the temporal bone of the cranium, whose name is in turn derived from the fact that the hair usually begins to turn grey over this area first, and thus marks the passage of time.

Attachments

- The temporal fossa.
- Fibres converge medial to the zygomatic arch to insert on to the coronoid process and ramus of the mandible.

Nerve supply

Anterior branch of the mandibular division of cranial nerve V (trigeminal).

Functions

- The anterior and superior fibres elevate the mandible and occlude the teeth.
- The posterior fibres retract the mandible and assist mastication.

Study tasks

- Draw lines carefully from the posterior part of the temporal fossa so that they insert on to the coronoid process parallel to the body of the mandible. Ensure that the anterior and superior fibres have a more vertical approach.
- With the direction of the fibres carefully drawn, consider the different functions of the posterior and anterior fibres.
- Palpate your own temporal fossa as you contract and relax temporalis muscle by clenching and relaxing your jaw. Test the different roles of the fibres by alternately occluding and retracting the mandible.

Protrusion

The mandible is drawn forwards in this movement so that the lower teeth move forward over the upper teeth. It is achieved essentially by the lateral pterygoids with assistance from the medial pterygoids and the superficial fibres of masseter. As a bilateral movement it has no particular use except that it produces a crudely quizzical facial expression. Unilaterally, it has great importance when combined with retraction on the opposite side, as it allows mastication to take place (p. 224).

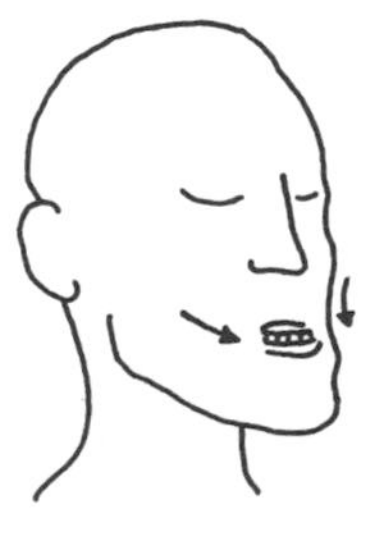

- lateral pterygoid (p. 217)
- medial pterygoid (p. 222)
- masseter (superficial) (p. 221)

Study tasks

- Highlight the names of the muscles shown.
- Study the details of each muscle with reference to the page number given.

- temporalis
  (posterior fibres) (pp. 222–223)

assisted by:
- masseter
  (middle and deep fibres) (p. 221)
- geniohyoid (p. 219)
- digastric (p. 218)

Retraction

The mandible is drawn back to its position of rest by contraction of the posterior fibres of temporalis, and more forcefully by geniohyoid, digastric and the deeper fibres of masseter.

Study tasks

- Highlight the names of the muscles shown.
- Study the details of each muscle with reference to the page number given.

Lateral movements (mastication)

Lateral movements are in fact an alternating 'side to side' tendency utilized in mastication. What actually happens is that the mandible protrudes on one side while it retracts on the other. Thus the lateral and medial pterygoids contract on one side (assisted by the superficial fibres of masseter) to produce unilateral protrusion, while the posterior fibres of temporalis contract on the opposite side to produce contralateral retraction with closure. This alternating pattern of movement combined with elevation and depression of the mandible produces the combined action of mastication.

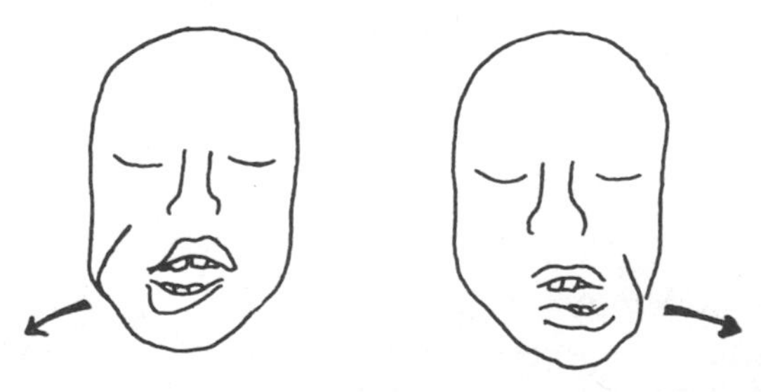

- medial pterygoid (p. 222)
- lateral pterygoid (p. 217)

} on each side, acting alternately with pterygoids on the opposite side

assisted by:
- temporalis (posterior fibres) (pp. 222–223)
- masseter (p. 221)

Study tasks

- Highlight the names of the muscles.
- Study the details of each muscle with reference to the page number given.

Glossary of terms and abbreviations

Most of the terms relevant to muscles and movement are explained in the introductory chapters (Chapters 2 and 3) with the aid of diagrams. The names of the individual bones and ligaments appear in the introductory diagrams in each chapter.

A

Abduction Movement of the body parts away from the midline.

Accessory process A small bony prominence at the root of the transverse process of a lumbar vertebra.

Adduction Return movement of the body parts towards the midline.

Anatomical position The universal position of the body accepted as a starting point of reference for anatomical description. It assumes that the body is standing facing the observer with palms facing anteriorly, legs together, and arms at the sides. All movements are described from this position.

Anatomical snuffbox The small hollow formed by the extensor and abductor tendons of the thumb in the extended hand. Snuff (tobacco) can be taken from this point. It is not recommended.

Angle Change in direction of bone, as in ribs, mandible, scapula.

Aponeurosis A flat sheet of tendon connecting one muscle to another, or to bone.

Apophyseal (facet) joints The synovial joints between the articular processes of the vertebrae.

Appendicular The peripheral parts of the skeleton including the limbs and limb girdles.

Arcuate Arched.

Articulation A joint, or movement between two or more bones.

Atypical Not typical, as in 'atypical synovial joint'.

Axial The central parts of the skeleton including the vertebral column, cranial bones, sternum and ribs.

Axis An imaginary line about which the body or a body part moves or rotates.

B

Belly With reference to muscle, it means the fleshy part of a muscle.

Bi- 'Two' or 'both'.

Bi-axial Two axes of movement.

Bifurcate To split into two parts.

Bilateral (bilaterally) On both sides.

Bursa A pouch of synovial membrane, usually located near joint surfaces in order to minimize friction.

C

C(1–7) Denotes cervical vertebrae, of which there are seven. The numbers may refer to either the root value of the cervical nerves (of which there are eight) or to individual vertebral segments, depending on the context.

Cardiovascular Relating to heart and circulation.

Carpus (Carpal) The collective term for the wrist.

Cartilage The connective tissue which covers the articular surfaces of bones, or joins bones together. There are two types: hyaline and fibrocartilage, and they also provide the constituent material for certain body parts such as the auricle (pinna) of the ear, intervertebral discs and menisci.

Cartilaginous joint Two articulating bones connected by either hyaline cartilage (a primary cartiaginous joint or synchondrosis) or by fibrocartilage (a secondary cartilaginous joint or symphysis).

Capsule The ligamentous fibres forming a casing around synovial joints. It is lined by synovial membrane.

Cervical Relating to the neck.

Circumduction A movement of the body or part of the body, in which the distal part describes the arc of a circle while the proximal attachment remains stable. It combines elements of extension, abduction, flexion and adduction.

Collateral At the sides.

Condyle A large articular prominence on a bone.

Condylar (bicondylar) joint Uni-axial synovial joint which nevertheless permits a degree of rotation, and is formed between the rounded convex condyle(s) of one articular bone and the concave accepting surface of the other.

Convergent Approaching a focal point from a more dispersed (divergent) starting point.

Compound As applied to joints, possessing more than one pair of articulating surfaces.
Concave A hollow shape.
Connective tissue Abundant tissue which supports, binds and gives physical integrity to the structures of the body.
Convex A rounded shape.
Contralateral On the opposite side.
Contraction With reference to muscle, the shortening action of the sliding filaments which produces movement.
Costal Relating to the ribs.
Cranium (cranial) The skull; relating to the skull.
Crest A prominent ridge, or border on a bone.
Cruciate Crossing, or cross-like.
Crus (crura) Support(s).

D

Deep Below a more superficial structure.
Depression Movement which lowers a part of the body. Hollow.
Distal Further from a point of attachment which is closer to the midline of the body.
Dorsal expansion A small aponeurosis covering the dorsal surface of the proximal phalanges of both the hands and feet, and providing important muscle attachments.
Dorsum (dorsal) The back or posterior surface.
Dorsiflexion Flexion of the foot at the ankle joint in an upward direction.
Dura mater The outer membrane covering the brain and spinal cord.

E

Elevation Movement which raises part of the body.
Ellipsoid joint A bi-axial synovial joint which has oval-shaped articular surfaces allowing flexion/extension and abduction/adduction.
Endomysium The protective cover surrounding a muscle fibre.
Epi- Above or outer.
Epicondyle A prominence above the condyle on a bone.
Epicondylitis Inflammation at the site of an epicondyle.
Epimysium The protective cover surrounding the belly of a muscle.
Eversion The outward movement of the sole of the foot.
Extension An increase in the anterior angle between articulating surfaces, which brings posterior surfaces closer together (except in the case of the knee, ankle and toes).
Extensor expansion See 'Dorsal expansion' above.

F

Facet A smooth, flat, well-demarcated surface on a bone.
Fascia Strong fibrous tissue which forms compartments between muscle. Its function is still not fully understood, but it should not be regarded as inert or totally non-contractile.

Fascicle (fasciculus, fasciculi) Bundle(s) of muscle fibres enclosed by perimysium.

Fibrocartilage The fibrous cartilage which is the main constituent of certain body parts, such as menisci and intervertebral discs. In some joints it covers the articular surfaces between bones instead of hyaline cartilage, but this is not typical.

Flexion The folding movement of a joint which brings the anterior surfaces closer together and decreases the anterior angle between the bones (except in the case of the knee, ankle and toes).

Foramen (foramina) An opening through which neurovascular structures may pass.

Fossa A depression on a bone or tissue surface.

Fusiform Lozenge-shaped.

G

Gliding The adjustive sliding movement between synovial plane joints, such as the intercarpal and tarsal bones of the wrist and foot.

Glottis The air passage between the vocal folds in the larynx.

H

Hinge joint A uni-axial synovial joint where the concave/convex articulating surfaces permit only flexion/extension.

Head The term is applied to muscle and bone. With reference to muscle, it means one of two or more proximal attachments to a bone. With reference to bone, it is the rounded articular part at one end, supported by a narrower shaft.

Hyaline cartilage A pliable type of cartilage, almost translucent in appearance, which covers the articular surfaces of most synovial joints, and also forms costal (rib) cartilage.

Hyper- Above.

Hypo- Below.

I

Infra- Beneath.

Inferior Towards the lower part of a structure; in a downward direction.

Inguinal Relating to the groin area.

Insertion With reference to muscles, the attachment point which moves the most, and is drawn towards the less mobile attachment point known as the origin, when a muscle contracts.

Inter- Between.

Interosseous Between bones.

Intra- Within.

Intrinsic Inside.

Inversion The inward movement of the sole of the foot.

Ipsilateral On the same side.

K

Kyphosis The primary flexion curve of the thoracic spine which, if exaggerated, produces a stooped or rounded back.

L

L(1–5) Denotes lumbar vertebrae, of which there are five. The numbers may refer to either the root value of the lumbar nerves or to individual vertebral segments, depending on the context.

L/S The lumbosacral disc area, which may also be designated L5/S1.

Labrum Rim.

Lateral Farther away from the midline of the body, or a structure.

Ligament Dense fibrous tissue that connects bone to bone.

Line With reference to a bone, a slightly raised ridge.

Linea aspera 'Roughened line' (latin translation).

Longitudinal Lying lengthways: it often refers to an axis of movement which runs along the shaft of a long bone, or from superior to inferior through the body.

Lordosis The secondary extension curve of the lumbar spine which, if exaggerated, produces an arched back and often protuberant abdomen.

Lumbar Relating to the low back, and specifically the vertebral column at L1–5 level.

M

Mamillary process A rounded rough prominence found on the posterior border of the superior articular processes of lumbar vertebrae.

Mastication Chewing.

Medial Closer to the midline of the body or a structure.

Median plane The imaginary flat surface that runs through the midline or middle, and divides the body into two equal left and right halves.

Mediastinum Partition between the lungs.

Meniscus (menisci) Crescent- or moon-shaped fibrocartilage disc(s) found within synovial joints.

Meta- Denotes a state or area of change.

Metabolic The biochemical changes that occur in the body.

Modus operandi The way in which something works.

Mortise (and tenon) A carpentry term to describe a piece of wood which has been precisely cut or drilled to accept another piece.

Multi- Many.

Multi- (poly) axial More than two axes of movement.

Myo- Referring to muscle.

Myofibril A bundle of myofilaments.

Myofilament The smallest thread-like structure found in striated skeletal muscle and forming an integral part of the 'sliding filament' mechanism.

N

Neck With reference to a bone, the narrow part supporting the head.

Neuro- Referring to nerves.

Nuchae (ligamentum nuchae) Of the 'nape' of the neck.

O

Oblique At an angle.

Occlusion The act of shutting or closing.

Opposition Flexion of the thumb towards the flexed fingertips, particularly between the thumb and little finger.

Origin With reference to muscles, the more stable attachment point which draws the more mobile 'insertion' towards itself when a muscle contracts.

P

Palmar Referring to the palm or anterior surface of the hand.

Palpation (palpate) Feel or gently touch for the purposes of evaluation or diagnosis.

Pennate Feathery in appearance.

Perimysium The covering which holds bundles of muscle fibres (fasciculi) together.

Phalanx (phalanges) Small bone(s) of the fingers and toes.

Pivot joint A uni-axial synovial joint which only permits rotatory types of movement.

Plane An imaginary flat surface that divides the body into sections.

Plane joint A synovial joint in which the surfaces are either flat or slightly curved permitting only sliding or gliding movements.

Plantar The undersurface of the foot.

Plantar flexion The downward movement of the foot at the ankle joint which 'points' the toes.

Posterior Towards or at the back of the body.

Process A prominent projection on a bone.

Pronation The rotatory movement of the forearm which allows the palm of the hand to face posteriorly. In the foot the equivalent is eversion, but is not produced by rotation.

Prone Lying horizontally, face downwards.

Protraction The movement of the scapula (shoulder blade) or mandible (lower jaw) in an anterior direction.

Proximal Nearer to a point of attachment which is closer to the midline of the body.

Q

Quadrate Square.

Quadrilateral Four-sided, but with only two parallel sides.

R

Ramus (rami) Root(s).

Raphe With reference to muscle, a tendinous seam between muscle tissue.

Retraction The movement of the scapula (shoulder blade) or mandible (lower jaw) in a posterior direction.

Retro- Lying behind.

Rhombus (rhomboid) Four equal sides, parallel but slanted, unlike a square.

Rotation Movement around a longitudinal axis in either an inward direction (medial rotation) or an outward direction (lateral rotation). Right or left in the vertebral column.

S

S(1–5) Denotes the sacrum which has five fused segments. The numbers may refer to either the root values of the sacral nerves or to the individual sacral segments, depending on the context.

Saddle (sellar) joint A bi-axial synovial joint in which the articular surfaces are concave in one direction and convex in the other, as seen on a saddle. This allows flexion/extension and abduction/adduction movements, and a very limited amount of rotation.

SP (SPs) An abbreviation denoting the spinous process(es) of individual vertebrae.

Sesamoid A relatively small bone found within a tendon (the name means 'seed-like'). Its purpose is to increase leverage, or alter the line of muscular pull.

Shaft The narrow tubular part of a long bone.

Sliding filament mechanism/theory The most widely accepted physiological explanation of how and why muscles contract. Consult a standard physiology text for further details.

Spine A sharp slender projection of bone.

Spheroidal joint (ball-and-socket joint) A multi-axial synovial joint between the ball-shaped head of a bone and the socket-shaped surface of another. It allows all ranges of movement.

Striated Striped.

Sulcus With reference to bones, a groove or depression.

Superficial At or relatively near the surface.

Superior Towards the upper part of a structure; in an upward direction.

Supination The rotatory movement of the forearm which allows the palm of the hand to face anteriorly. In the foot the equivalent is inversion, but is not produced by rotation.

Supine Lying horizontally, face upwards.

Supra- Above.

Symphysis A secondary cartilaginous joint in which the articulating bones are held together by fibrocartilage.

Syndesmosis A joint in which the articular surfaces are held together by strong fibrous tissue, allowing only strictly limited movement.

Synergist An assisting muscle which has a stabilizing effect and reduces undesirable movement.

Synovium (synovial) The membrane which lines joint capsules and secretes a lubricating fluid.

T

T(1–12): Denotes thoracic vertebrae, of which there are twelve. The numbers may refer to either the root value of the thoracic nerves or to individual vertebral segments, depending on the context.

TP (TPs) An abbreviation denoting the transverse process(es) of individual vertebrae.

Tarsus (Tarsal) A collective term for the posterior part of the foot.

Tendon Dense white connective tissue which joins the muscle to bone.

Thenar Refers to the palm of the hand.

Thorax (thoracic) The chest; 'thoracic' may refer to the middle part of the vertebral column from T1 to T12 where the ribs attach.

Thoracolumbar A junction area between the thoracic and lumbar regions.

Tonic With reference to muscle, it describes the normal resting state.

Torso The trunk, or that part of the body excluding head and limbs.

Trabeculae A lattice of thin lines showing patterns of force in spongy bone.

Transverse Across, or at right angles to the length of a structure.

Trapez- Suggests a four-sided figure with only one pair of parallel sides (from 'trapezium').

Trochanter A large blunt projection found near the head of the femur (thigh bone).

Tubercle A small rounded projection on a bone.

Tuberosity A large rounded, often roughened projection on a bone.

UVW

Uni- 'One' or single.

Uni-axial One axis of movement only.

Unilateral On one side only.

Unciform/uncinate A hook-shaped lip on the superior and lateral border of a cervical vertebral body.

Valgus Displacement or angulation away from the midline.

Varus Displacement of angulation towards the midline.

XYZ

Yomping Forced marching as carried out by the soldiers of the Royal Marines.

References

Dowdney, G. (1993) More than skin deep: reflections from the dissecting room. *Journal of Osteopathic Education*, 3(2), 122–7.

Grant's Atlas of Anatomy (1991) Williams and Wilkins, Baltimore.

Gray's Anatomy (1995) 38th edn, Churchill Livingstone, Edinburgh.

Hartman, L.S. (1990) *Handbook of Osteopathic Technique*, Chapman & Hall, London.

Kapandji, I.A. (1987) *The Physiology of the Joints*, Churchill Livingstone, Edinburgh.

Snell, R.S. (1995) *Clinical Anatomy for Medical Students*, 5th edn, Little Brown and Co., Boston.

Index

Page numbers appearing in *italic* refer to figures. Note: where the term (circumduction) appears in parentheses it requires a Study Task response in the text. Explanations of the meaning of this, or any other, term should be sought in the *Glossary of terms and abbreviations*.

Abdominal breathing 169
Abdominal muscles 182–6
  see also individually named muscles
Abduction, *see appropriate individual joints (movements)*
  see also named abductor muscles; Interossei, dorsal
Abductor
  digiti minimi (foot) 164
  digiti minimi (hand) 92
  hallucis 164
  pollicis brevis 81
  pollicis longus 80
Adduction, *see appropriate individual joints (movements)*
  see also named adductor muscles; Interossei, palmar and plantar
Adductor
  brevis 116
  hallucis 165
  longus 115
  magnus 116
  pollicis 82
Agonist 12
Anatomical 'snuffbox' 77
Anconeus 52
Angle of inclination (of hip) 98, *98*
  of torsion (of hip) 98, *98*
Ankle
  bones *141*
  joints *142*
  ligaments *142–3*
  movements
    dorsiflexion 143, 144
    plantar flexion 143, *147*
  muscles, *see specific ankle movements*
  see also individually named muscles
Antagonist 12
Aponeurosis, definition of 10
  see also Muscle(s), attachments
Arches
  of foot *154*
  of hand 73, *73*
Articularis genus *128*
Attachments, *see* Muscle(s), attachments

Back muscles, *see* Vertebral column, muscles
Biceps
  brachii 34, 49
  femoris 109
Bones, *see individual joints*
Brachialis 48
Brachioradialis 50
Breathing, *see* Respiratory muscles

Calf muscles, *see* Ankle, movements, plantar flexion
Carpal bones and joints, *see* Wrist
Carpal tunnel 60
Carrying angle (of elbow) 47, *47*
Cervical spine
  bones (vertebrae) *188*
  movements
    circumduction 188
    extension *193*
    flexion *190*
    rotation 212, *213*
    sidebending (lateral flexion) 212, *213*
  muscles, *see specific cervical movements*
  see also individually named muscles
Chewing, *see* Mastication
Circumduction, *see appropriate joints (movements)*
Coccygeus 123
Coracobrachialis 33–4
Cubital, *see* Elbow

Deep breathing, *see* Respiratory muscles, accessory
Deltoid 32
Diaphragm
  pelvic 121
  thoracic 168–9
Digastric 218
Dorsiflexion, *see* Ankle, movements

Elbow
  bones *45*
  joints 45–7, *46, 53*
  ligaments *46, 53*
  movements
    extension 51, *51*
    flexion 48, *48*
    pronation 53–4, *56*
    supination 53–4, *54*
  muscles, *see specific elbow movements*
  see also individually named muscles

Erector spinae (sacrospinalis) 197–203
see also individually named muscles
Eversion, *see* Foot, movements
Expiration, *see* Respiratory muscles
Extension, *see appropriate individual joints (movements)*
see also named extensor muscles
Extensor
carpi radialis brevis 67
carpi radialis longus 66
carpi ulnaris 68
digiti minimi 91
digitorum 90
digitorum brevis 163
digitorum longus 145
hallucis brevis, *see* digitorum brevis
hallucis longus 146
indicis 91
pollicis brevis 79
pollicis longus 78
External intercostal muscles, *see* Intercostalis(-es), externi
External oblique muscle, *see* Obliquus, externus abdominis

Fascia lata 114
Fixator 12
Flexion, *see appropriate individual joints (movements)*
see also named flexor muscles
Flexor
carpi radialis 62
carpi ulnaris 63
digiti minimi brevis (of foot) 162
digiti minimi brevis (of hand) 86
digitorum accessorius 160
digitorum brevis 158
digitorum longus 151
digitorum profundus 85
digitorum superficialis (sublimis) 84
hallucis brevis 161
hallucis longus 150
pollicis brevis 75
pollicis longus 77
Flexor retinaculum (of wrist) 60
Foot
arches *154*
bones *154*
joints 153, *154*, 157
ligaments *154*
movements
abduction (of toes) 163, *163*
adduction (of toes) 165, *165*
eversion (pronation) 153, *155*
extension (of toes) 157, *162*
flexion (of toes) 157, *157–8*
inversion (supination) 153, *155*
muscles, *see specific foot movements*
see also individually named muscles

Gastrocnemius 147–8
Gemellus (the Gemelli)
inferior 119–20
superior 119
Geniohyoid 219–20
Gluteus
maximus 108–9
medius 112–13
minimus 113
Gracilis 105–6

Hamstring muscles 109–11
see also individually named muscles
Hand
arches 73, *73*
bones *73*
see also Wrist, bones
joints 72–4, *73–4*
ligaments 74, *74*
movements (fingers)
abduction *72, 92*
adduction *72, 93*
circumduction, *see* Wrist
extension *72, 89*
flexion *72*, 83, *83*
opposition *72*, 93, *93*
movements (thumb)
abduction *72, 80*
adduction *72, 81*
circumduction 83, *83*
extension *72, 77*
flexion *72, 75*
opposition *72, 82*
muscles, *see specific hand movements*
see also individually named muscles
Hip
angle of inclination 98, *98*
angle of torsion 98, *98*
bones 95–8
joints 95–8
ligaments *99*
movements
abduction 112, *112*
adduction 114, *114*
(circumduction) 121
extension 107, *107*
flexion 100, *101*
lateral (external) rotation 117, *117*
medial (internal) rotation 117, *117*
muscles
superficial *100*
see also individually named muscles; specific hip movements

Iliacus 103
Iliocostalis
cervicis 201–2
lumborum 203–4
thoracis 202–3
Iliocostocervicalis, *see* Iliocostalis
Iliopsoas, *see* Iliacus; Psoas, major
Iliotibial tract (band) *100*, 114
Infrahyoid muscles 220
Infraspinatus 42
Insertion, of muscles, *see* Muscle(s), attachments
Inspiration, *see* Respiratory muscles
Intercostal muscles, *see* Intercostalis (-es)
Intercostalis (-es)
externi (pl.) 172–3, *172*
interni (pl.) 179–80, *180*
intimus 179–80, *180*
Interossei
dorsal (foot) 160
dorsal (hand) 87, 88
palmar 89
plantar 159
Interspinales 209
Intertransversarii 208
Inversion, *see* Foot, movements

Jaw, *see* Temporomandibular joint
Joints
ankle 141–51
cartilaginous *15*
classification 15–17
elbow 45–57
fibrocartilaginous (secondary cartilaginous/symphysis) *15*
fibrous *15*
finger 73–4

foot *154*
hand 71–94
hip 95–121
knee 125–39
pelvis 96–9; 121–3
sacro-iliac *96*, 97
synovial *16–17*
temporomandibular 215–25
thumb 72, 74–83
toes 157–65
wrist 59–69

Knee
bones 125–8, *126*
joints 125–8, *127–8*
ligaments *127*
menisci 126, *127*, 130–1
movements
extension 132–3, *133*
flexion 131–2, *132*
lateral (external) rotation *139*
medial (internal) rotation *138*
muscles
superficial *129*
see also individually named muscles; specific knee movements

Latissimus dorsi, 37, 176
Levator
ani 122
costarum (levatores) 169–70
scapulae 26
Levers 18–19, *18*
Ligaments, *see individual joints*
Longissimus
capitis 200
cervicis 200
thoracis 201
Longus
capitis 191
colli 190–1
Lumbricals
foot 158–9
hand 86–7

Masseter 221
Mastication 224
Movements, *see individual joints*
Multifidus 205–6
Muscle(s)
arrangement (architecture) *13–14*
attachments 10, *10*
derivation of names 19
insertions, *see* attachments
origins, *see* attachments
pennate (types of) *14*
skeletal, voluntary striated 9
superficial
of abdomen *182*
of hip *100*
of knee *129*
of shoulder *23*
of vertebral column (back) *194*
teamwork 12
testing, *see under various Study Tasks with reference to individually named muscles*
Mylohyoid 218–19

Naming of muscles, *see* Muscle(s), derivation of names
Nerve supply of muscles, *see individually named muscles*

Obliquus
capitis inferior 211
capitis superior 210–11
externus abdominis 184
internus abdominis 184
Obturator
externus 118–19
internus 118
Omohyoid 220
Opponens
digiti minimi (of hand) 94
pollicis 76
Opposition, *see* Hand, movements (thumb)
see also individually named muscles
Origin of muscles, *see* Muscle(s), attachments

Palm of hand, *see* Hand
Palmaris
brevis 94
longus 64
Pectineus 103
Pectoralis
major 32–3, 174
minor 29, 175
Pelvic diaphragm 121–3
Pelvis
bones *96–7*
joints *96*, 97
ligaments *99*
movements 97
see also Pelvic diaphragm
muscles, *see* Hip
see also Pelvic diaphragm
Peroneus
brevis 156–7
longus 156
tertius 146
Piriformis 121
Plantar flexion, *see* Ankle, movements
Plantaris 149
Popliteus 138–9
Power, of muscle 13
Prime movers 12
Pronation, *see* Elbow, movements
Pronator
quadratus 56
teres 57
Psoas
major 102
minor 193
Pterygoid
lateral 217
medial 222
Pyramidalis 183

Quadratus
femoris 120
lumborum 178
plantae, *see* Flexor, digitorum accessorius
Quadriceps femoris 134–7
see also individually named muscles
Quiet breathing, *see* Respiratory muscles (primary)

Range, of muscle movement 173
Raphe, definition of 10
Rectus
abdominis 183
capitis anterior 192
capitis lateralis 208–9
capitis posterior major 209–10
capitis posterior minor 210
femoris 104–5, 134
Rectus sheath *183*
Respiratory muscles
expiration
accessory muscles (deep breathing) 167, *179*
primary muscles (quiet breathing) 167, *179*
inspiration
accessory muscles (deep breathing) 167, *167*
primary muscles (quiet breathing) 167, *167*
see also individually named muscles

Rhomboideus (the rhomboids)
  major 29–30
  minor 29–30
Rotation, *see appropriate individual joints (movements)*
  see also named rotator muscles
Rotator cuff, of shoulder 22, 23–4
Rotatores 204–5

Sacro-iliac joint *96*, 97
Sacrospinalis (-es), *see* Erector spinae
Sartorius 105
Scalenus
  anterior 170
  medius 171
  posterior 171–2
Scapular movements 24–5
Semimembranosus 111
Semispinalis
  capitis 206
  cervicis 206
  thoracis 207
Semitendinosus 110
Serratus
  anterior 28, 177
  posterior inferior 181
  posterior superior 178
Shoulder
  bones 21, *21*
  joints 21, *22*
  ligaments *22*
  movements
    abduction 38–9, *38*
    adduction 40, *40*
    circumduction 43
    extension 35, *35*
    flexion 30–1, *31*
    lateral rotation 41, *41*
    medial rotation 40, *40*
    scapular 24–30
  muscles
    superficial 23
    see also individually named muscles; specific shoulder movements
Soleus 148
Spinalis
  capitis 198
  cervicis 198
  thoracis 199
Splenius
  capitis 195–6
  cervicis 196
Sternocleidomastoid 173
Sternocostalis, *see* Transversus, thoracis
Sternohyoid 220
Sternomastoid, *see* Sternocleidomastoid
Stylohyoid 220
Subacromial surface 39–40
Subclavius 43
Subcostales 180
Suboccipital muscles 209–11
  see also individually named muscles
Subscapularis 41
Superficial muscles, *see* Muscle(s), superficial
Supination, *see* Elbow, movements
Supinator 55
Suprahyoid muscles 218–20
  see also individually named muscles
Supraspinatus 39
Synergists 12

Tarsal bones and joints, *see* Foot
Temporalis 222–3
Temporomandibular joint (TMJ)
  bones *215*
  ligaments *216*
  movements
    depression (opening) 216, *217*
    elevation 220, *220*
    lateral movements (mastication) 224
    protrusion 223–4
    retraction 224
  muscles, *see specific TMJ movements*
  see also individually named muscles
Tensor fasciae latae 114
Tendons 10, *10*
Teres
  major 36
  minor 42
Thumb, *see* Hand
Thyrohyoid 220
Tibialis
  anterior 144
  posterior 149–50
Toes, *see* Foot
Transversospinalis muscles 204
Transversus
  abdominis 185
  thoracis 180–1
Trapezius 27–8
Triceps brachii 51–2

Valsalva manoeuvre 169, 185
Vastus
  intermedius 136
  lateralis 137
  medialis 135
Vertebrae, *see* Vertebral column
Vertebral column
  bones (vertebrae) 187–8, *188*
  movements
    circumduction 188–9
    extension *189*, 193
    flexion *189*, 190, 192
    rotation *189*, 212, *213*
    sidebending (lateral flexion) *189*, 212, *213*
  muscles
    deep 197
    deepest (intrinsic) 204
    intermediate 194–5, *195*
    superficial 194, *194*
  see also individually named muscles; specific vertebral movements; Abdominal muscles; Hip movements, flexion; Respiratory muscles

Wrist
  bones 59, *59*
  joints 59–61, *59*
  ligaments 60–1, *60*
  movements
    abduction *68*
    adduction *69*
    (circumduction) 69
    extension 65, *65*
    flexion 61, *61*
  muscles, *see specific wrist movements*
    see also individually named muscles

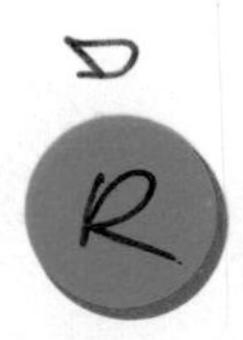
R